The Roman Land Surveyors

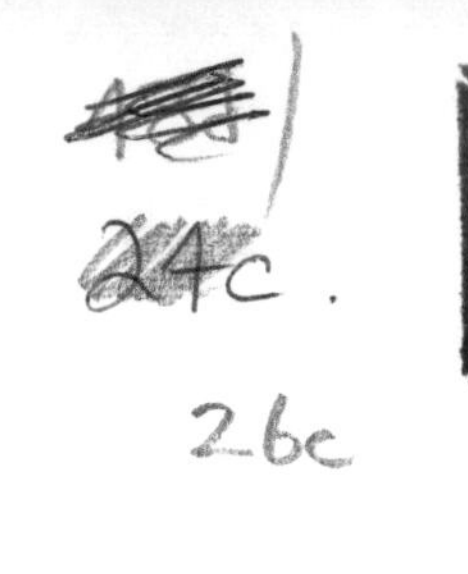

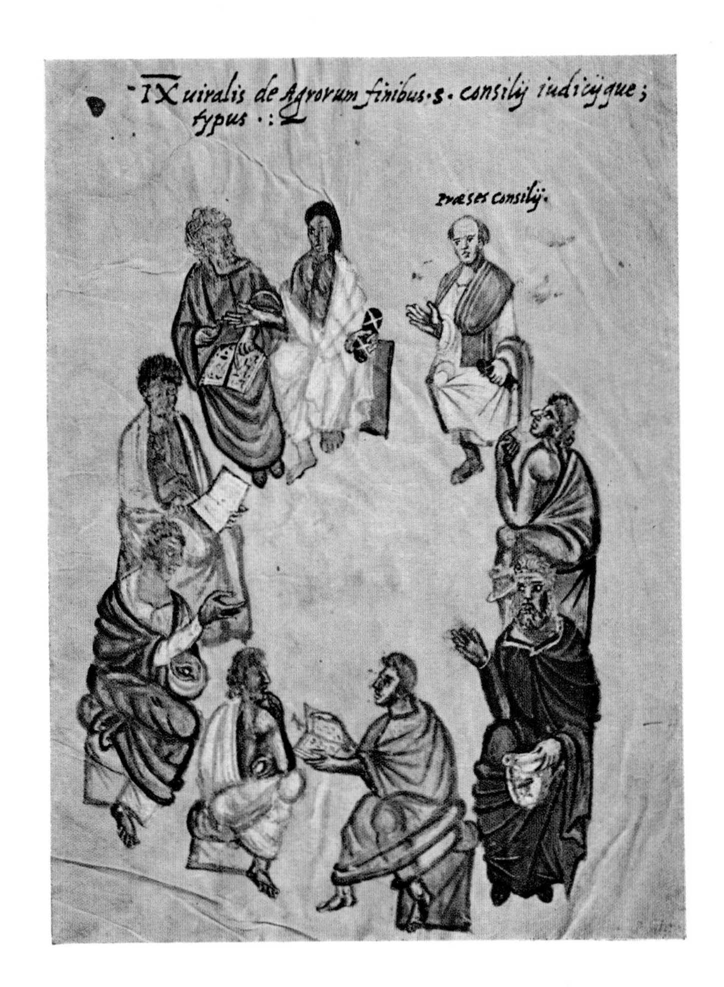

Land Commission in session under the Empire

THE ROMAN LAND SURVEYORS

An Introduction to the Agrimensores

O. A. W. DILKE

Professor of Latin in The University of Leeds

DAVID & CHARLES: NEWTON ABBOT

ISBN 0 7153 5279 2

Set in Monotype Bembo
and printed in Great Britain
by Latimer Trend & Company Limited Plymouth
for David & Charles (Publishers) Limited
South Devon House Newton Abbot Devon

Contents

		Page
1	Introducing the *agrimensores*	15
2	Pre-Roman surveying and geodesy The Near East—Egypt—Greece and Greek Colonies	19
3	Roman land surveying and surveyors The early period—the Etruscans—the Republic—the early Empire—the late Empire	31
4	The training of Roman land surveyors General education—geometry of areas—orientation—sighting and levelling—distance calculation—astronomy and cosmology—the law and the surveyor	47
5	Roman surveying instruments *Groma* and *stella*—portable sundial—miscellaneous equipment—*chorobates*—*dioptra*—hodometer	66
6	Measurement and allocation of land Roman surveying measurements—orientation in practice—division of land—allocation of land	82
7	Boundaries Roman boundaries and the land surveyor—boundary disputes	98
8	Maps and mapping General maps—surveyors' working maps—surveyors' teaching maps	109
9	Roman surveying manuals The *Corpus Agrimensorum*—its illustrations—related works	126

Page

10 Centuriation 133
Origins and causes
I. The study of Roman surveyed land
Work on centuriation—unorthodox systems
II. Centuriated areas
Italy—Gaul and adjacent areas—Dalmatia—Greece—Asia Minor and adjacent areas—North Africa

11 The Orange cadasters 159
The setting—the three cadasters—summary

12 Colonies and State domains 178
Colonies and *ager publicus* under the Republic—colonies under the Empire—the *libri coloniarum*

13 Accurate planning in Roman Britain 188
Centuriation and parcelling—town plans—roads—the Fenland—the Antonine Wall distance slabs

14 Roman surveying and today 201

Finding list 212

Glossary 214

Notes and references 218

Appendix A Contents of the *Corpus* 227

B *Kardo* and *decumanus* 231

Bibliography 234

Acknowledgements 251

Index 253

List of Illustrations

Land commission under the Empire. From Codex Palatinus Vaticanus Latinus 1564, Biblioteca Apostolica Vaticana (=MS P) *frontispiece*

PLATES

	Page
Fresco from tomb at Thebes, Upper Egypt. From Sir Henry Lyons in *Geographical Journal* 69 (1927). Photograph Science Museum, London	49
Merkhet (Egyptian surveying instrument). From Sir Henry Lyons, as above. Crown copyright. Science Museum, London	49
Surveyor's cross from the Fayyum. From Sir Henry Lyons, as above. Crown copyright, Science Museum, London	49
Tombstone of L. Aebutius Faustus. From H. Schöne in *Jahrbuch des kais. deutsch. archäol. Inst.* 16 (1901)	50
Reconstruction of *groma* found at Pompeii. From M. Della Corte in *Monumenti Antichi* 28 (1922)	50
Roman surveying instrument found at Pfünz, Bavaria. From H. Schöne, as above	67
Iron end-pieces of a wooden measuring rod. From E. Nowotny, *Römische Forschung in Oesterreich, 1912–1924*	67
Seated geometer or surveyor. From Codex Guelferbytanus 36.23 Aug. 2° (Arcerianus), Herzog August Bibliothek, Wolfenbüttel (=MS A)	67
Kardo maximus and *decumanus maximus* faultily surveyed. From MS A, Wolfenbüttel	68
Admedera (Haïdra, Tunisia). From MS P, Vatican	68

Page

Colonia by the sea. From MS P, Vatican 68

Upper miniature: unsurveyed land. Lower miniature: *subseciva*. From MS A, Wolfenbüttel 101

Area of *subseciva* near the River Pisaurus. From MS A, Wolfenbüttel 101

Boundary lines and triangular boundary mark. From MS A, Wolfenbüttel 102

Boundaries between adjacent farms. From MS A, Wolfenbüttel 102

Fragments of the Forma Urbis Romae. Capitoline Museum, Rome 119

Two fragments of the Arausio (Orange) cadasters. Musée Archéologique, Orange (many fragments destroyed in a collapse) 119

Anxur-Tarracina (Terracina). From MS P, Vatican 120

Air photograph of Terracina, taken by the Royal Air Force. British School in Rome 120

Minturnae. From MS P, Vatican 153

Minturnae. Photograph by the author 153

Hispellum (Spello). From MS P, Vatican 154

Centuriation round a settlement. From MS A, Wolfenbüttel 154

Mt Massicus and Suessa Aurunca. From MS A, Wolfenbüttel 171

Augusta Taurinorum (Turin), Hasta (Asti) and other places. From MS P, Vatican 171

Augusta Praetoria (Aosta). From MS P, Vatican 172

Colonia Claudia, presumably Aventicum (Avenches). On same page as Augusta Praetoria 172

Common pasture land. From MS A, Wolfenbüttel 172

Page

Inscription of Arausio (Orange) setting up a cadaster. From A. Piganiol, *Les documents cadastraux de la colonie romaine d'Orange* 189

Air photograph of part of the Rhône valley, with reconstruction of the centuriation. From A. Piganiol, as above 190

Air photograph of the area between Hedjez el Bab, Oudna and Grombalia, Tunisia. Supplied by Professor R. Chevallier 190

Air photograph of part of the Campanian plain, taken by the Royal Air Force. British School in Rome Collection 207

Air photograph of central Naples, taken by the Royal Air Force. British School in Rome Collection 208

IN THE TEXT

1 A *groma* (based on the Pompeii *groma*, p 50) 16

2 Miletus. From A. von Gerkan, *Griechische Städteanlagen*, with later corrections by the same writer 24

3 Columella. Area of rectangular field 52

4 Columella. Area of trapezoidal field 53

5 Columella. Area of segmental field 53

6 Columella. Area of hexagonal field 55

7 Nipsus. Finding perpendicular and base of a right-angled triangle 55

8 Nipsus. A similar exercise 56

9 Hyginus Gromaticus, orientation. From MS A (Wolfenbüttel) 57

10 Hyginus Gromaticus. Exercise in parallel lines. From MS A (Wolfenbüttel) 59

11 Nipsus. Measuring width of a river. From MS A (Wolfenbüttel) 60

Page

12 Reconstruction of Nipsus's diagram. From Hyginus, *Liber de munitionibus castrorum*, ed C. C. L. Lange (1848) 61

13 Nipsus. Cross-check on survey. From MS A (Wolfenbüttel) 61

14 Hyginus Gromaticus. Orbits round the earth. From MS A (Wolfenbüttel) 62

15 Hyginus Gromaticus. Plan with legal definition of status of an area. From MS P (Vatican) 64

16 Ivory box from Pompeii serving as portable sundial and measure. From M. Della Corte, 'Groma', *Monumenti Antichi* 28 (1922) 71

17 Portable sundial from Crêt-Chatelard. H. Diels, *Antike Technik* 72

18 Reconstruction of *chorobates* 74

19 Reconstruction of the *dioptra* of Hero of Alexandria. From Hero, ed H. Schöne (Teubner) 75

20 Reconstruction of sighting device on measuring rod designed for use with Hero's *dioptra*. From Schöne (above) 77

21 Reconstruction of Hero's leveller. From Schöne (above) 78

22 Reconstruction of the mechanism of Hero's hodometer. From Schöne (above) 80

23 Hyginus Gromaticus. Centuriated areas which have rectangles (a) 40 × 20 *actus* (b) 20 × 40 *actus*. From MS A (Wolfenbüttel) 83

24 Hyginus Gromaticus. Centuriation stone. From MS A (Wolfenbüttel) 89

25 Latin abbreviations for the four main areas of a centuriation system 90

26 Hyginus Gromaticus. The four central 'centuries'. From MS A (Wolfenbüttel) 91

Page

27 Method of numbering 'centuries' 92

28 Hyginus Gromaticus. Centuriation showing *quintarius*. From MS A (Wolfenbüttel) 93

29 Frontinus. (a) *Strigae* and *scamna*, (b) private land. From MS A (Wolfenbüttel) 95

30 Hyginus Gromaticus. Differently aligned systems of centuriation. From MS A (Wolfenbüttel) 100

31 Latinus. Boundary stone. From MS G (Wolfenbüttel) 104

32 Notitia Dignitatum. Rough map of Britain. From O. Seeck's edition 110

33 Notitia Dignitatum. Map of Egypt. From O. Seeck's edition 111

34 Tarracina-Anxur. Remains of centuriation. From *Greece and Rome* n.s. 8 (1961) 117

35 The area SW of Hispellum 121

36 The keeping open of *limites*: Frontinus 124

37 The fallacy of taking bearings by sunrise or sunset. From MS A (Wolfenbüttel) 127

38 Limburg province. Ancient land division. From C. H. Edelman and B. E. P. Eeuwens, *Berichten van de rijksdienst voor het oudheidkundig bodemonderzoek* 9 (1959) 140

39 Map of centuriation sites in Italy. From F. Castagnoli, *Le ricerche sui resti della centuriazione* 143

40 Centuriation in part of the Po valley. From R. Chevallier, *Caesarodunum* Suppl 2 (1967) 145

41 Modern roads near Lugo, Po valley, preserving pattern of centuriation 147

42 Map of centuriation sites in Tunisia. Based on maps by R. Chevallier 152

Page

43 Centuriation and ancient land divisions near Enfida, Tunisia. Adapted from *Atlas des centuriations romaines de Tunisie* 155

44 Fragment 7 of Orange Cadaster A. From A. Piganiol, *Les documents cadastraux de la colonie romaine d'Orange* 164

45 Orange–Avignon area. Remains of centuriation. Based on air survey by M. Guy (published by R. Chevallier in *Caesarodunum*) 165

46 Orange Cadaster B, IV F. Reconstruction. From A. Piganiol, ibid 167

47 Orange Cadaster B, III C 171. From A. Piganiol, ibid 168

48 Orange Cadaster B, 193–196. From A. Piganiol, ibid 170

49 Orange Cadaster C, 351–357. From A. Piganiol, ibid 174

50 Roads near Cliffe, north of Rochester. From M. D. Nightingale, *Archaeologia Cantiana* 65 (1952) 192

51 Ancient land divisions near Ripe, Sussex. From I. D. Margary, *Sussex Archaeological Collections* 81 (1940) 194

52 Silchester. Outline of Roman town and roads. Adapted from G. C. Boon, *Archaeologia* 102 (1969) 197

53 USA: problem of converging meridians. From W. D. Pattison, *The Beginnings of the American Rectangular Land Survey System* 206

VXORI DILECTISSIMAE

1

Introducing the agrimensores

Agrimensores, 'measurers of land', were the land surveyors of ancient Rome. But they not only measured it: they laid it out with more careful planning and more accuracy than in any country at any time until the late eighteenth century.

Many ancient civilisations, from early Babylonia and dynastic Egypt onwards, had officials who measured land. Boundary marks, in many countries protected by religious taboos, became a prime concern of surveyors: in the Commination, in the Book of Common Prayer, we find:

> Cursed is he that removeth his neighbour's land-mark.
> Amen.

The Romans were a practical and methodical people. For surveying land they used an instrument called the *groma*, a cross mounted (in the specimen found at Pompeii, Fig 1) on a bracket, which fitted into a staff. It was essentially designed to survey straight lines and right angles, and so Roman land divisions are regularly in squares or rectangles. The square was by far the commoner.

The surveyor of state lands which were to be allocated would start at a chosen point and plot a *limes*, or dividing line, in each of the four directions planned, which often corresponded approximately to the four cardinal points. The squares which these delimited, each often with sides of 2,400 Roman feet (770–6yd, 705–10m) and containing an area of 200 *iugera* (ca 124·6 acres/50·4 hectares), were called *centuriae*, 'centuries', because they theoretically contained 100 plots of the early size of a smallholding. This division of land into 'centuries' is known as centuriation, and was particularly applied to colonies, settlements of smallholders with allocations of land. As the Roman sphere of influence spread, there were more public domains to be divided up and more new colonies were founded. The land surveyors were in great demand.

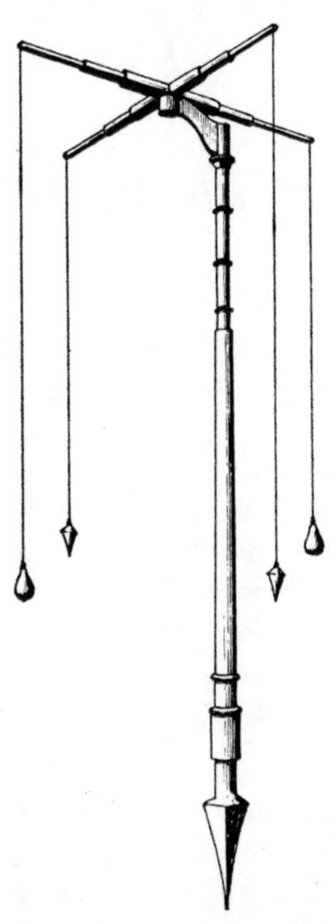

1. A *groma* (based on the Pompeii *groma*, p 50)

A regular training of surveyors was organised under the Empire. It included cosmology and astronomy, the geometry of areas, orientation, sighting and levelling, a knowledge of land law and of the status of different types of land, as well as the techniques of centuriation, boundary definition, allocation of land, mapping and recording. What training there was for military and architectural surveyors was, it seems, separately organised. Under the late Empire the whole structure became more bureaucratic, and the *agrimensores* rose in status. They became also judges or arbitrators in cases where land law wasi nvolved.

The Corpus Agrimensorum is a collection of surveyors' manuals which has come down to us in often corrupt and fragmentary texts. They are preserved in manuscripts of which the most important, at Wolfenbüttel and in the Vatican, date from the sixth and ninth centuries. Two of these contain many miniatures whose colouring is very well preserved. Although the maps in these miniatures are copies several times removed, we can recognise in a number of them interesting topographical features. The more detailed plans in the Corpus tend to serve as teaching models, but working plans in the heyday of Roman surveying must have contained much accurate recording. The monks who copied the collection were unaccustomed to the difficult technical Latin and probably understood little of the details of texts and diagrams. There were only a few scholars, like Abbot Gerbert, later Pope Silvester II, who were interested in the subject in the Dark and Middle Ages. Only in very recent times has any idea of the importance of the miniatures to the history of cartography been disseminated. Earlier works on Roman cartography gave as specimens mainly the distorted Tabula Peutingeriana, the impressive Forma Urbis Romae, and the world maps based on Ptolemy, who was not a Roman but an Alexandrian Greek.

The standard edition of the Corpus is *Die Schriften der römischen Feldmesser*, or *Gromatici Veteres*, in two volumes originally published by F. Blume and others in 1848 and 1852. The first volume contains the Latin text, edited by Lachmann (hence, eg, 15L is Vol 1 of the Corpus, page 15) and the diagrams. The second volume has essays and useful indices. The principal writings of the Corpus were also edited by C. Thulin in the Teubner series, with small black-and-white photographs of the MS miniatures. The earliest technical writer in the Corpus is Sextus Julius Frontinus, governor of Britain probably from AD 74 to 78, author of works on strategy and on Rome's water supply. Among the other writers is one who shows acquaintance with Roman campaigns on the northern frontier.

Our other chief source of information about details of centuriation came mainly after World War II from a large number of inscribed stones found at Orange, in Provence. They belong to three cadasters, detailed land surveys for taxation purposes, and show the main intersections and rivers as well as abbreviated particulars of holdings in each 'century'.

Much research has recently been concentrated on the remains of Roman land division. Many areas of centuriation[1] have been found in Italy and Tunisia, pronounced remains in Yugoslavia and Southern France, and traces of centuriated land elsewhere. The most extensive areas are in the Po valley and Tunisia. The former was systematically colonised with Roman settlers, especially in the first half of the second century BC. The centuriation there, planned on a very large scale, has left conspicuous remains. In North Africa, most of the land round Carthage became *ager publicus* (State domains) after the defeat of Carthage in 146 BC, and centuriation in the northern area followed. Very large tracts of southern Tunisia were centuriated, at least in skeleton, in the first century AD.

In most places the network of squares of land and *limites* is perfectly regular. But in some we find, or read of, various irregularities: squares of irregular size, rectangles, strip systems. In the areas of Europe more distant from Italy, including Britain, there is as yet little evidence of regular centuriation. With the archaeological and photographic resources at our disposal, however, much more may well be discovered.

The activities of the *agrimensores* continued not only to the end of the Western Empire but well beyond. Although in the next chapter what is known of pre-Roman and Ptolemaic surveying is outlined, none of the other ancient peoples had the same opportunity or quite the same thoroughness as the Romans. The Roman Empire extended so widely and lasted so long that it prompted careful organisation and left many permanent records. Its administrators were thorough: land had to be measured accurately, allotted appropriately and recorded fully and permanently.

2

Pre-Roman surveying and geodesy

In the earliest civilisations of the Near East we find many indications that land was regulated. In Babylonia, votive stones called *kudurus* were placed by landowners on their fields. On each of these they recorded the exact boundaries of the field and invoked the curses of a god on any who violated these boundaries. A letter from Hammurabi on a boundary dispute has been interpreted in two different ways: whereas the nineteenth-century editor[1] thought that the king ordered the dispute to be settled by a land survey, others have thought that he left it to the judgement of the sun-god. Disputes about irrigation were common in Hammurabi's reign, and we also find mathematical problems concocted by military engineers.[2] The Babylonians and Egyptians knew that a triangle with sides of 3, 4, and 5 units has a right angle. The Babylonians used a sexagesimal notation, eg 70 was expressed as 1,10. Neugebauer[3] shows that their equivalent for $\sqrt{2}$ was 1·414213, a remarkably close approximation to the true 1·414214. . . .

Apart from the well-known Babylonian world-map, we have certain more detailed maps from the early kingdoms. Thus an Egyptian map of ca 1300 BC known as the Turin papyrus[4] shows mountains in which there were gold and other mines, and the roads leading from these to the Red Sea. This unique papyrus has been called not only one of the first topographical maps but even the first geological map, since it colours the mountains near the gold-mine pink, while other mountains are coloured black. The gold-mine has plausibly been located at Umm Fawakhîr in the Wadi Hammamât.

A tablet from Tello, Mesopotamia, now in the Istanbul Museum, has the properly dimensioned plan of a royal estate amounting to about

515 acres/208ha. A cuneiform tablet of ca 1300 BC from Nippur, Mesopotamia, gives a plan of nearby fields and canals, with six townships and royal land indicated. It is interesting to find that irrigation in these areas led to the development of mathematics, although some of the problems set were rather academic. One of these concerns the depth of water (from a cistern of given dimensions) with which a plot of land about 3 miles (4·8km) square would be flooded, the answer being a mere finger's breadth.

Egypt, to the Greeks who visited it, was the land of the pyramids and of exact geometrical calculations. Moreover they rightly recognised the antiquity of Egyptian civilisation and institutions. Although they did not always hit upon the right explanation of the flooding of the Nile, they could clearly see what an important part it had played in land survey and in the application of geometry to this.

Herodotus,[5] after mentioning the Egyptian canal system originated by Sesostris, continues:

> The priests also said that this king divided the country among all the Egyptians, giving each an equal square plot. This was the source of his revenue, as he made them pay a fixed annual tax. If some of anyone's land was taken away by the river, he came to the king and told him what had happened. Then the king sent men to look at the land and measure how much less it was, so that in future the owner would pay the due proportion. It seems to me that land survey started from this and passed on to Greece. The concave sundial and the division of the day into twelve were learnt by the Greeks from the Babylonians.

In fact the plots were not all square and equal, and the part said to have been played personally by the Pharaoh may well be exaggerated. In Genesis 47: 20–6 we are told that during the famine Joseph bought all the land of Egypt for Pharaoh and made a law that one-fifth of the produce of it should be given to the Pharaoh, and that apart from priests' land this law still existed. Both Herodotus's account and that of the Bible (with the addition of military land) contain some truth. If Sesostris is the right name, the Pharaoh concerned could have been any of three successive ones, perhaps Sesostris II, 1897–1878 BC. But the whole practice is likely to have had an even earlier start.

The entire economy of Egypt was governed by the annual flooding of the Nile. After each flood, boundary marks might well be in different positions and boundary lines obliterated. Surveyors were therefore needed to work out new lines, and they worked for centuries on similar principles. The fame of Egyptian surveyors was still known to Cassiodorus in the sixth century AD. It was in the interest of authority as well as of individuals to have boundaries established, since taxation was involved. We learn from Egyptian records that from the second dynasty onwards the king had a census of landed property and other possessions carried out for taxation purposes. In the Theban empire taxation was based partly on an inventory of property which was frequently revised. Every year officials measured the arable land and listed its owners, making an estimate of the produce and hence of the likely tax yield. Later, during the growing season, officials, including a legal scribe, two scribes from the survey office, a cord-keeper and a cord-stretcher came round to each area to assess the amount of tax. The boundaries were inspected, and owners had in some cases to swear that the boundary-marks had not been moved.

Measurements of length in the agricultural communities of early Egypt were calculated from parts of the body. For the purpose of land measurement, various sizes of cubit (forearm) were used. The *merkhet*, [5a] or plum-line sighter (p 49), was in service from very early times. It consisted of a holder with a short plumb-line and plummet. The *merkhet* was aligned on an object by looking through the split centre-rib (held upwards) of a palm-leaf. Even in very early times the Egyptians recognised the importance of recording the flood levels of the Nile. The Palermo Stone, in the museum at Palermo, has a record of royal annals of the Old Kingdom with river gauge markings. From the second dynasty we find mention of a numbering of gold and lands, and a biennial census developed. In the third dynasty we hear of a property register which was kept in the royal archives. As taxes were proportionate to areas held, the latter were carefully recorded. Unfortunately we have no surviving map of such property.

With the help of a *merkhet*, of measuring rods and of a measuring cord of 100 cubits, the Egyptian surveyors were clearly able to achieve great accuracy. The easiest check on this is to assess the measurements of the Great Pyramid. Its four sides, intended as a square, at the base, of 440 Royal cubits, each of 52·35 or 52·36cm, vary only between 230·253 and 230·454m. The slope of the sides is only 6mm out from the level east–west, 14mm north–south. The mean axes are on no side more than 5′ 30″ from the true compass point. At the pyramid of Meidun,

Flinders Petrie found levelling-lines marked out at intervals of 1 cubit, to guide the masons.

Various tombs and statues give us an insight into surveying practice. The tomb of Mes at Saqqara reveals that he had to defend his title to lands given to an ancestor against a rival who in the law-court produced false title-deeds, so that the official registers had to be brought. The tomb of Menna at Sheikh Abd el Qurna, Thebes (p 49), is of particular interest. It shows two chainmen measuring a field of corn with a long cord which contains knots at intervals of about 4 to 5 cubits; each has a spare cord wound round his hand. As a peasant brings bread and corn, three scribes, whose writing materials are carried by a slave, are ready to record the particulars. The religious connection of surveying is brought out in the statue of Pa-en-hor from Abydos (now in the Cairo Museum) who holds a rolled measuring cord. Remains have been found of inscribed stelae which King Ikhnaton (fourteenth century BC) set up on both sides of the Nile valley; although not all were found, it is thought that the distances between pairs were extremely close to each other.

The study of archaic and classical Greece has so far revealed little evidence of systematic land surveying. The Greeks, as Herodotus shows, thought of themselves as having learnt the principles of surveying from Egypt. But the city-states of Greece had, for the most part, few wide expanses of flat arable land, and the cities tended to be founded round defensible sites which were not easy to divide into regularly shaped plots. The legacy of Egypt lay rather in geometry, used partly in its literal sense of measuring the earth (this now falls under geodesy) and partly in its Euclidean sense. Thales of Miletus and the other philosophers of the Ionian school were the chief representatives of this science in the eastern half of the Greek world, while the Pythagoreans worked on it in the West. Pythagoras, who emigrated from Samos to Croton (Cotrone, S. Italy) ca 531 BC, may well have been the first to prove the theorem which bears his name. Archytas of Tarentum was most famous for his discoveries in acoustic and geometric theory; but Horace, interpreting the Greek word *geometres,* calls him a 'measurer of land and sea and of the infinite particles of sand'.[6] Related activities in Palestine and Greece included tunnelling, in which a good degree of accuracy was reached. Polycrates of Samos (tyrant ca 540–522 BC) employed Eupalinus of Megara to build a tunnel which can still be seen, about 1km long, 1m 75 high and wide.

The practical activities of Greek planners—for the Greeks were not only theorists—were more geared to the planning of cities[7] than of the countryside. The profession concerned with the regular planning of cities was the architect's. Whereas the traditional layout of a Greek city was rather lacking in planning, there are a number of cases where a regular pattern is observable. When the colony of Olbia, on the northern shore of the Black Sea, suffered a fire at the end of the sixth century BC, part of it was rebuilt with a rectangular pattern. Although there are some parallels to be found in the near East, it is probable that the Greeks of Miletus, who founded Olbia as one of their many maritime colonies, thought out this plan independently. The most famous name associated with square and rectangular town-planning schemes was also a man of Miletus, Hippodamus.[8] We are told that after the Persian wars he redesigned the Piraeus on a grid pattern; the dating has been confirmed by an examination of the boundary stones, the work having evidently been instigated by Themistocles. It must have been Hippodamus who, when his own city of Miletus was rebuilt in 479 BC, designed that too on a system of squares, following in general the lie of the land on the peninsula (Fig 2). In 443 BC he took part in the foundation of the Greek colony at Thurii in South Italy; we are told that its original streets numbered only four one way and three the other. Whereas many Greek cities preferred their traditional irregular layout, in some his pattern was carried on after his death; at Priene, Olynthus and elsewhere we can see examples of this. Priene incorporated in its squares and rectangles all the public and private buildings neatly grouped round a central market place. Olynthus had an old and a new sector: in the old there was no discernible pattern, whereas the new was based on groups of two squares each forming rectangles surrounded by streets. When the 'new town' of Megalopolis was founded about 371 BC, its buildings were on both sides of a river, but there was a common orientation to the two sides. Many of the Hellenistic cities continued this type of regularity, though some purposely departed from the rigid straight lines. We are told by Strabo that at Nicaea in Asia Minor the gymnasium was planned centrally so as to give a view down four streets, at the end of which one would see the four main gates of the city; such an arrangement was most unusual.

Many of these activities involved only architects in their planning.

But when a new colony was founded, the founder sometimes took with him not only priests, to establish the cult of the patron god or goddess and duly inaugurate the colony, but surveyors. It was the latter's function to divide up the land of the new colony, since, although the

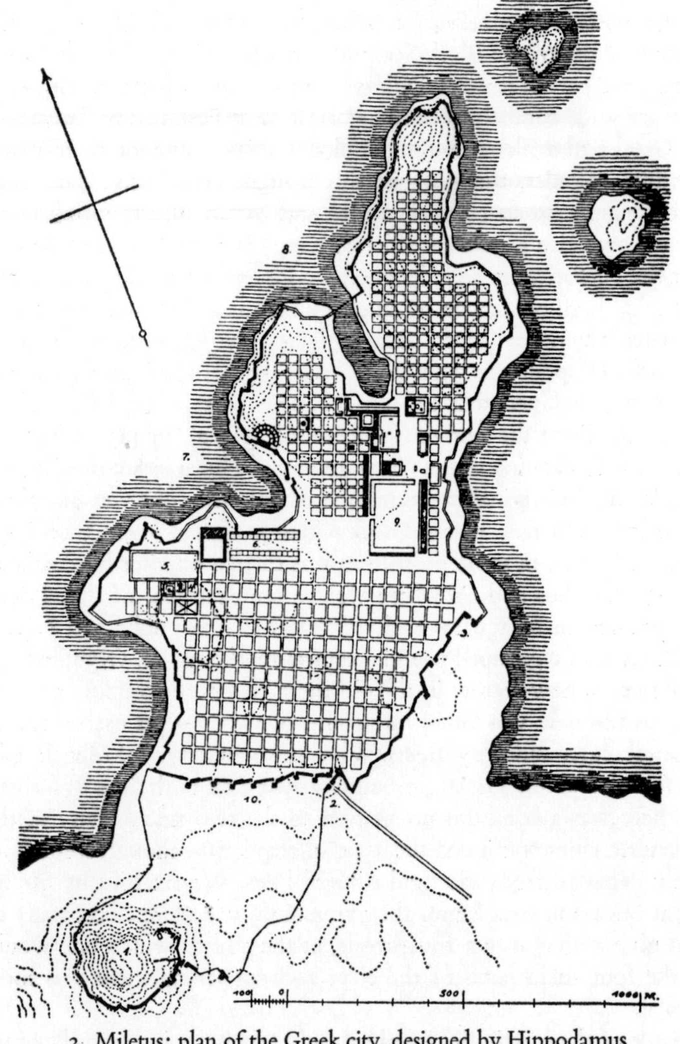

2. Miletus: plan of the Greek city, designed by Hippodamus

Greeks expanded mainly to promote their maritime trade, most of their colonies had a fair amount of arable land. Hence we should expect it to be more in these colonies than in the mother cities that Greek town and country planning showed a correlation. The evidence, while scanty, certainly tends to confirm this. At Selinus (Selinunte, Sicily), where the south part of the town was restored after being sacked by the Carthaginians in 409 BC, the street system bears some resemblance to the Roman *kardines* and *decumani*. Some of the Greek colonies in Southern Italy have been investigated since World War II with the help of aerial photography. As a result, something of the pattern has begun to emerge, especially on the Gulf of Taranto and at Caulonia (near Monasterace Marina). This is a valuable addition to our knowledge of pre-Roman surveying, since in most other areas of the classical Greek world anything in the nature of a rectilinear pattern tends to be urban, not rural planning, though the squares and rectangles of Greek cities may well have influenced the shapes of Roman land divisions. But in the Greek colonies of Southern Italy, as we shall see, urban and rural planning were to some extent combined.

Between the Greek colonies of Heraclea and Metapontum (Metaponto) on the Gulf of Taranto there has been accretion of the coastline: the prehistoric coast was 3 to 4km farther north-west than the modern, the early classical coastline about 1km farther. The land systems which emerge from aerial survey are mostly inland from the prehistoric coastline. In this area two different Greek land orientations are observable, roughly ENE and NNE respectively. Each consists of a series of very long strips, about 230 to 240m wide. One is visible for about 6km, one for 7km. The more easterly system of the two has the same orientation as the streets of Metapontum and clearly belonged to that colony. The orientation of the western system is largely dictated by the approximately parallel courses of the Torrente Cavone and the R. Basento, between which it lies. To which Greek colony it belonged is uncertain. Results from Caulonia show similar patterns; and sites in Corsica, on the Adriatic and on the Black Sea have also been examined. The claim[9] that a square of 200m is visible from aerial photography of Agathe Tyche (Agde, Hérault) seems to the present writer doubtful.

Evidence for the existence of land registers in classical Greece is very thin. The lexicographer Harpocration, writing under the Roman

Empire, says that in Athens of the classical period the demarchs (chief officials of the demes, ie townships) had copies made of the lands in each deme. Plato[10] speaks of publishing records of sales contracts so as to ensure that no injustices are done, but does not mention anything like a cadaster.

The Alexandrian and other Hellenistic Greek scientists paid particular attention, among other subjects, to geodesy. In the period of Alexander's conquests the knowledge of distant countries increased. Thus Pytheas of Marseilles circumnavigated Britain towards the end of the fourth century BC, and Megasthenes, ca 295 BC, reliably described the Ganges valley. With a long period of comparative peace in the Greek world, this type of geographical knowledge was used to help scientists work out the measurement of the earth. Dicaearchus, who lived mostly in the Peloponnese, established ca 310 BC a basic line of latitude which extended from the Straits of Gibraltar to the Himalayas. Euclid of Alexandria (ca 300 BC) wrote textbooks of geometry which lasted to the twentieth century and had a profound impact on the theory of Roman surveying. Eratosthenes (ca 275–194 BC) was the greatest mathematical geographer of antiquity. His lines of latitude and longitude intersected at Rhodes, and his greatest achievement was the calculation of the earth's circumference. Many Greek scientists upheld the sphericity of the earth against popular flat-earth tradition. Eratosthenes's calculations were based on his knowledge of Egyptian geography. A well was dug at Syene (Aswan) to the bottom of which the sun penetrated at midday on the summer solstice, and Eratosthenes reckoned that it must be on the tropic. He worked out the distance from Syene to Alexandria as 5,000 stades, and the difference in angle of the sun at Alexandria as one-fiftieth of a circle, which gave him $50 \times 5{,}000 = 250{,}000$ stades (later modified to 252,000) as the circumference of the earth. He checked at the winter solstice and found the result the same. In fact Aswan is about 60km north of the tropic and 3° east of Alexandria, but even with these inexactitudes he obtained a very close approximation, closest to the true measurement if he used a short stade of 157·5m. Hipparchus the astronomer, making observations at Rhodes and elsewhere between 161 and 126 BC, divided the main parallel of latitude set up by Eratosthenes into 360° and propounded a method of fixing locations by latitude and longitude.

A story about Alexander the Great shows that he had absorbed from his tutor Aristotle some sound ideas about town and country planning. Dinocrates wanted to carve Mount Athos in the likeness of Alexander's face and hands, the left hand holding a large city and the right a water-basin. 'No,' said Alexander. 'Just as a baby cannot nourish itself and grow without its nurse's milk, neither can a city without fields and produce flowing into its walls.' So he made Dinocrates plan the new city of Alexandria with harbour, city and surrounding fields as a single unit.

Egypt under the Ptolemies has been shown by papyrus and other discoveries to have had a most elaborate land system. This was geared to taxation, which distinguished between inundated and non-inundated land.[11] The latter was taxed at least as high, probably higher, as indeed we should expect, since the farmers near the Nile were deprived of the use of their fields during the flooding. In this period fuller use of irrigation opened up new lands to cultivation, including some in the Fayyûm.

Excavations in the north of the Fayyûm in 1899 unearthed a simple surveyor's cross (p 49), perhaps of the Ptolemaic period. It was intended to be held in the hand, and was made of two pieces of palm-leaf rib 35·2 and 34·2cm long, tied together at right angles with palm fibre cord. The upper rod has an incision to keep the cross at the correct angle. Originally attached to the ends, but not discovered by the excavators, were plumb-lines with plummets; each arm had a deep notch near the end to receive these.

Papyri have revealed many details of land use in Egypt during the Ptolemaic period. Every year the village clerk (*komatogrammateus*) of each village had to draw up a report on the ownership and cultivation of lands in his district, so as to help the government assess the revenue. The papers of Menches, village clerk of Kerkeosiris near Tebtunis ca 120–111 BC, and his successor Petesouchus (110–108 BC) give much interesting detail of land registration. The Tebtunis area, round Umm el Baragât in the south of the Fayyûm, had during the previous hundred years or so been reclaimed as agricultural land and settled with soldiers and others. Following the general pattern, the village clerk of Kerkeosiris drew up annual reports showing who owned lands in the area, what crops he grew, and what revenue was obtained from them. In the year 119/18 BC the total was 4,700 *arourai*, about 3,200ac/1,295ha (the Greek word *aroura* literally means 'arable land', but as a Ptolemaic unit of area it denoted 100 cubits square), divided into the following seven categories:[12]

Class of land	*Area in* arourai
Local authority land	69½
Unproductive, untaxed	169 9/16
Religious	271⅞
Smallholdings	1,564 27/32
Gardens	21¼
Pastures, untaxed	175⅜
Royal land	2,427 19/32

As pointed out by the editors of the papyri, the subdivision is inexact, since it depends partly on ownership and partly on nature of land, with some overlap. Also it is likely that, despite the apparently careful fractions, one at least of the figures is 'cooked', ie arrived at by subtraction from the probably round estimate of 4,700.

The breakdown of smallholdings for the year 120/19 BC gives the following table, which classifies the holders and shows in which reign the original assignments of land to each category of men were made:

No.	*Category*	*Ptolemy IV*	*Ptolemy V*	*Ptolemy VI*	*Ptolemy VIII*	*Total* (*in* arourai)
29	military colonists	70	114⅜	378⅞	402	965¼
1	mounted desert guard	34 3/32	—	—	—	34 3/32
3	desert policemen	—	10	20	—	30
3	police officials	—	—	—	—	30
2	inspectors	—	—	—	48	48
8	cavalry of Chomēnis	—	—	—	120	120
55	seven-*aroura* fighting cavalry	—	—	—	354	354
	Totals	104 3/32	124⅜	398⅞	924	1,581 11/32

It will be seen that the holdings varied, according to status of settler, from 6½ to over 34 *arourai* for the categories other than military colonists—in fact the seven-*aroura* cavalry received only 6½ *arourai*, all except one who had part of his holding in the territory of an adjacent village. For the military colonists the largest holding mentioned in the area (not in this particular list) is 100 *arourai*. Land is classified as arable, pasture and pasture-grass, desert, saltmarsh, and flooded. The crops on the arable land are listed (wheat, barley, lentils, wild chickling, beans, peas, mustard etc), the farmer, if not identical with the landowner, is mentioned by name, and the tax on each plot is given. Mistakes in arithmetic are not uncommon. There is in this particular area no question of annual flooding by the Nile.

Correspondence from a tax-farmer (i.58) mentions a land-surveyor (*geometres*) of Thebes in Upper Egypt, Acusilaus son of Paos. This was clearly an important official, who had issued a memorandum affecting a number of village clerks as well as the writer. The name Acusilaus is Greek, but the father's name is not. In addition we hear of an official, Theon (i.24), who was appointed to survey the vineyards and gardens. The same letter complains of the number of men who have wormed their way into various bureaucratic jobs; evidently surveys were carried out by very many differently named civil servants. Although this was a defect of the central administration, it was not so in the village. A letter from Petesuchus, Menches's successor, reads:

> List resulting from the enquiry held by Horus the royal clerk after the same year's survey of the arable land in Kerkeosiris which is in the category of estate revenue, stating also the excess amounts which result. The total reclaimed from unprofitable land is 78 *arourai*, of which 16½ *arourai* are under cultivation . . . I swear by Queen Cleopatra and King Ptolemy, by the deified Ptolemies, by Serapis and Isis and all the other gods and goddesses that I have presented the above mentioned report and have made no false statement. If my oath is true, may it be well with me; if not, the reverse.

Clearly, therefore, the local survey or much of it was done by the village clerk.

Another Tebtunis papyrus tells us that on the east side of Crocodilopolis Arsinoe (Medinet el Fayyûm), between the road round the town and a canal, there were many small plots, whose dimensions are given in *schoenia* (100 cubits); some were private, one belonged to the queen, one was common land. The Tebtunis papyri also include a few land documents from Magdola.

Oxyrrhyncus Papyrus 918, of the second century BC,[13] also comes from the Fayyûm. It concerns Crown land at the village of Ibion

Argaei, especially marshy areas let out by the treasury. An entry[14] reads: 'South of this, rented in the third year (Ptolemy unspecified), by A[.]apeos son of Heron and the other village elders, 18¾ *arourai*, the flooded ones, taxation reduced in the fourth year. This includes the 6¾ *arourai* which the survey of the land bordering the water, carried out in the eleventh year, found to be cultivated and no longer flooded. Total, as above, [18¾].'

The method of calculating irregular quadrilaterals in Egypt was rough and ready: it was to multiply the averages of opposite sides. Thus in the quadrilateral below, a rough area would be obtained by the formula $\frac{1}{2}(AB + CD) \times \frac{1}{2}(AD + BC)$. In the case of a convex quadrilateral this always results in an over-estimate, which as Déléage[15] points out benefited the treasury. The method, however, persisted, and we find it in the Corpus. Certain documents give the lengths of sides of fields, but without more evidence we cannot with any confidence reconstruct a map of them.

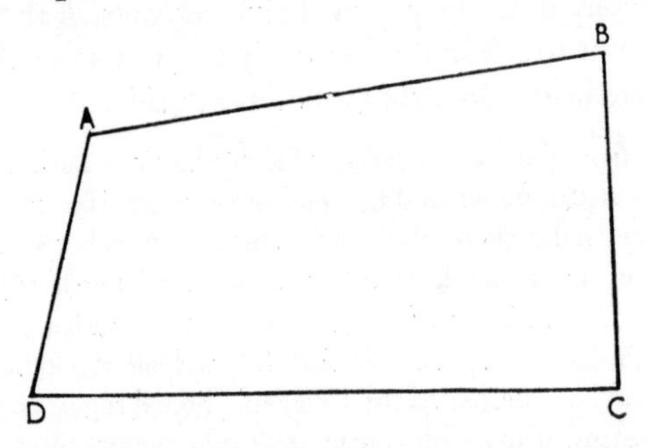

It has been thought[16] that land contracts under the Seleucid dynasty in Syria indicate that there must have been survey maps, but this is uncertain. In one of these contracts, dated 254/3 BC, the archivist Timoxenus is ordered to enter a sale, with a note of the boundaries, in the royal ledgers at Sardis.

This outline has brought us to the end of the second century BC in the Hellenistic world, when Roman surveying had been operating for a considerable time. It had learnt much, in one way or another, from the predecessors of Rome. But in an area like Egypt it did not in its turn influence local practice. In civilised parts of the Roman world, where there was an established routine, Rome usually left it undisturbed.

3

Roman land surveying and surveyors

The Romans, in early days a nation of farmers, attributed great antiquity to land surveying. When Virgil gives a picture of Aeneas founding a city in Sicily, he uses the words:

> *interea Aeneas urbem designat aratro*
> *sortiturque domos*[1]
>
> Meanwhile Aeneas marks the city out
> By ploughing; then he draws the homes by lot

This refers to the old custom by which the consul marked the limits of a new town by ploughing a furrow round it. He held the plough obliquely, so that the earth fell inwards, and lifted the plough over areas intended as gateways. He wore his toga hitched up in the *cinctus Gabinus*, a method of wearing it derived from the old town of Gabii, 19km east of Rome. Ovid[2] ascribes the dividing up of land with balks (*limites*) by a 'careful measurer' (*cautus mensor*) to the last of his four ages, the Iron Age, whose people were destroyed by the Flood. In his youth Ovid had studied law, and he must at least have touched on the legal background to surveying. The choice of epithet and the juxtaposition of *limite* and *mensor* point to careful phraseology, and the insertion of surveying in his brief 'history of civilisation' shows the importance he attached to it.

The technical phrase for the line drawn round a town, as alluded to by Virgil, was *sulcus primigenius*, 'the original furrow'. According to Tacitus and Plutarch this constituted the outer boundary of a newly founded city, being marked with boundary stones. This tallies with boundary stones found at Capua (S. Maria di Capua), originating in the

Second Triumvirate. The inscription on these[3] may be rendered 'By order of Caesar (Octavian), on the line ploughed'. Similarly in the foundation charter of Julius Caesar's colony Colonia Iulia Genetiva Urbanorum at Urso (Osuna, Spain), preserved on bronze tablets in Madrid, it is stipulated that no one shall bring a corpse inside the territory of a town or colony, where a line has been drawn round with a plough. On the other hand the elder Cato, according to Isidore, spoke of actually building a wall where the founder had ploughed; and Ovid and other writers confirm this version. Ovid[4] almost echoes Virgil's words quoted above, but in speaking of Romulus founding Rome says he marked out the walls with a plough. As Thulin[5] observes, one can imagine that this piece of ritual originated with the wall circuit; it belongs in origin to lowland, and often had to be modified when applied to towns built on rocks. The area immediately outside the walls was called *pomerium*, and was not allowed to be inhabited or ploughed. Livy[6] and others consider the custom an Etruscan one.

How much of Roman surveying can be traced back to the Etruscans? Frontinus, in the Corpus, tells us that balks (*limites*) between plots originated, according to Varro, in the 'Etruscan lore' (*disciplina Etrusca*). Varro's *Antiquities* (47 BC) was a very learned work, but we know that he was not infrequently mistaken. Hence doubt has been thrown on this, as on other ancient statements, about Etruscan origins. Frontinus goes on to explain that soothsayers (*haruspices*) faced west and divided their observation zone into four parts, thus:

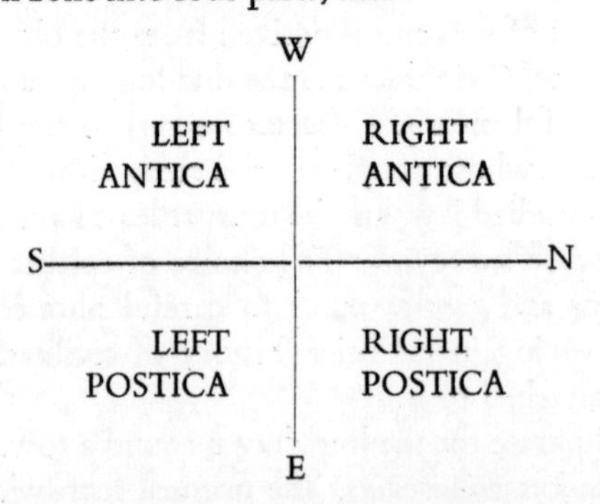

Since the terms *antica*, 'to the front', and *postica*, 'to the rear', were obsolete by the time of these writers, we cannot be sure that they hand down a correct tradition. Moreover Varro,[7] who gives the same terminology, makes the augurs face south. But if the terminology is correct, it certainly provides a close parallel to the surveyors' system, which substituted for these terms 'beyond the *kardo*' (p 90) and 'this side of the *kardo*'. The augural name for an augur's field of vision was *templum*,[8] which also meant 'temple'. Where religious rites were concerned, Rome undoubtedly owed much to Etruria; and some link between augury and surveying seems extremely likely. Moreover the word *groma* (see below and Chapter 5) was almost certainly derived from Greek through Etruscan.

Other evidence is less clear. An Etruscan bronze representation of a sheep's liver,[9] found near Piacenza in 1877, has a line dividing the zone of the sun from the zone of the moon. There is no doubt that it was used by soothsayers, and each zone is divided into sectors, triangles etc, with an Etruscan inscription in each. But to claim that it shows two main cross-divisions, a *decumanus* as well as a *kardo*, is fanciful. Its shape is irregular and dictated by the actual shape of a sheep's liver, and it does not have any intersection at right angles.

The Romans insisted on setting the *groma*, the cross-staff with which the bulk of their land division was carried out, with correct auspices. In Chapter 7 it will be seen that the whole notion of boundaries and boundary marks had a profound religious significance to them. It is also worth remarking that the word *pontifex*, which means a member of the college of priests concerned with public ritual, must in origin have meant 'bridge-maker'. In other words, bridge-building was regarded as needing some sort of magic powers. But we cannot be sure what Etruscan connection there was in all this.

Two small links with Etruria are worth mentioning. First, the Corpus preserves what purports to be an injunction by Vegoia, ie the Etruscan nymph Begoe, to the soothsayer Arruns that those who remove boundary stones will be cursed by the gods. This piece is shown by a reference to the avarice of the eighth century (after the foundation of Rome) to have originated in the first century BC.[10] It was probably devised as a piece of propaganda. Secondly, the word *acnua*, meaning the same as *actus quadratus*, a piece of land 120 Roman feet square, may perhaps be of Etruscan origin, although Columella records it from Spain: *acn-* is an Etruscan verbal root.

This is not, however, to say that the idea of centuriation in squares can be shown to be Etruscan. Excavations at Etruscan sites like Marzabotto have indeed revealed some sort of planning, but it is either rectangular or more haphazard. Similarly aerial photographs of Tarquinia and Spina[11] show rectangular or random patterns, not square ones.

To Carthage, as the great rival, Rome was in general unwilling to acknowledge any debt. It is, however, significant that the agricultural manual of Mago in twenty-eight books was translated both into Greek and, by order of the Roman senate, into Latin. This Latin translation appeared after the elder Cato's extant work on agriculture (ca 160 BC).

Rome was also indebted to the Greek colonies in South Italy, not merely for literature and art but for some of her ideas about the division of land. Investigations carried out recently on the Greek colonies of the Gulf of Taranto (p 23) and elsewhere have shown that parallel strips were an established land division in Italy long before the Roman supremacy.

It would be wrong to claim that the Romans simply copied the system which they found in the Greek colonies of South Italy and Sicily. Whereas the land patterns of these emerge as strips or rectangles, the characteristic shape of Roman centuriation is square. This, as has been seen, was the town-planning pattern introduced into Greek cities by Hippodamus of Miletus. What the Romans did was to combine features from Egypt, Etruria, Greek towns and Greek countryside to make their own distinctive system. Two of Rome's other debts to Greece lay in their numerical notation and their mathematical theory. Many must have wondered why, if M stands for *mille*, D (=500) does not correspond to any Roman equivalent. The answer is that the Romans borrowed a numerical system from the Chalcidian colonies of South Italy, in which 1,000 was written *Φ* or CIↃ, which became conflated with M in the Roman system. For 500 they bisected this first sign to become IↃ, which looks like D and was accordingly so written. The underlying geometrical theory came from Pythagoras, Euclid, Hero (under the Empire) and other Greek mathematicians; while for the more complicated surveying instruments they were likewise indebted to the Greeks.

In Chapter 12 the early development of Roman colonisation is outlined. Three colonies were even, without much support, said to have been planted as early as the period of the kings. Presumably the active initiators of the earliest settlements were the augurs, a college of official diviners whose duty it was to discover by various signs whether the gods approved or disapproved of a particular action. The number of augurs increased from three to sixteen. We do not know what part was played in planning these early settlements or their successors by trained or experienced surveyors. There is good evidence that the planning was organised carefully, the sites chosen with skill, the land properly centuriated and the plots methodically allocated. For the foundation of each colony in historical times a commission was established, usually of three men, often including ex-consuls. They were appointed for a specified period, eg three years. The convener of the commission was only exceptionally one of the consuls;[12] the chairmanship may have rotated from year to year. The duties of a commission were to define the boundaries of a new colony, assign allotments, settle any disputes, draw up the foundation charter and appoint the first magistrate and priests of the colony. Many names of land commissioners are preserved; they come from forty-seven *gentes*, enumerated by MacKendrick.[13] They were given *imperium*, so that they could take any military measures necessary or override local authorities. Among their staff were surveyors, but one has the impression that under the Republic these did a far smaller proportion of the work, and the members of the land commission a far larger, than under the Empire. As a sample of their activities, it is reckoned that between the years 200 and 190 BC a million *iugera* of land were distributed to 100,000 families.[14]

The relative importance of land commissioners as opposed to surveyors is to some extent confirmed from the nomenclature. The official title of a commission of three was *tresviri coloniae deducendae*, 'commission of three for founding a colony', literally leading it out, which indicates that in origin the settlers were led from Rome to their destination. To this phrase was sometimes added *agroque dividundo*, 'and for parcelling land'. The name given under the Republic to assistant commissioners was *finitores*, 'definers'. Plautus, in the prologue to his *Poenulus*, 48f, imagines the actor who is speaking it as a land commissioner or surveyor setting bounds to the plot:

> *eius nunc regiones, limites, confinia* | *determinabo; ei rei ego sum factus finitor.*
>
> I shall now delimit its areas, *limites* and boundaries; I have been appointed its surveyor.

In Rullus' agrarian law opposed by Cicero in his consulship, 63 BC, the

200 *finitores*[15] of equestrian rank were assistant land commissioners. According to Nicolet[16] they were sons or grandsons of Roman knights, and Rullus' purpose was to form a substantial body of respectable, well-educated and capable young men. The grammarian Nonius Marcellus, writing in the early fourth century AD, says that *finitores* were what had come to be called *agri mensores*. One may doubt if the equation is exact, since *mensor* or *metator* in the late Republic and early Empire is a humbler individual (usually freedman) than the equestrian land commissioners of Cicero's time; and even by the late Empire, although *mensores* had improved their status, it was still not uniformly high.

In the period from 133 BC, growth of population coupled with inequality of opportunity led under Tiberius and Gaius Gracchus to extensive redistribution of land, foundation of new (including controversial) colonies, and political rioting. For a reformer like Tiberius Gracchus to promote nepotism to such an extent as to appoint his father-in-law Appius Claudius, his brother Gaius and himself as the members of his land commission was political madness. In vast schemes like that of Junonia (Carthage; p 156), although it is clear that the land commissioners were active, they could not have carried out their task if they had not been provided with a fair number of surveyors.

Two early inscriptions show us different spheres in which their services must have been heavily relied on. From 117 BC we have the inscription recording the arbitration by Quintus and Marcus Minucius, which fixed the boundaries of the Langenses and the Viturii in North Italy (p 100).[17] The other record of this same decade which affects surveying is the Agrarian Law of 111 BC, commonly known as 'Lex Thoria'.[18] So much of it is preserved that it can only be very briefly summarised here. It provided, among many other things, that public land in Italy occupied since 133 BC and not exceeding 30 *iugera* should be considered private and rent-free; that ten head of cattle or so many smaller animals (number missing) could be grazed by anyone on common pasture lands; that lands given by land commissioners to a colony or an individual since 133 BC should be confirmed if necessary. With regard to land in Africa, public land, with certain exceptions, was to be sold by Roman magistrates; colonists could claim occupied land as private unless it exceeded 200 *iugera*, or unless the number of colonists exceeded the permitted maximum; various forms of compensation were payable, eg where the same land was sold to two persons; seven loyal towns (p 156) should be exempted from the land division and compensated; except for certain categories, all land in the province of Africa should be subject to rents, tithes, and pasture-tax. Land at Corinth was to be measured out and sold; but on this we have not the full details from the inscription.

The terms describing surveyors under the Gracchi are uncertain, but in the first century BC they were *metator* and *mensor*. Both of these mean 'measurer', but whereas Cicero uses the former (also *decempedator*, from the ten-foot rule), it was the latter that prevailed. Cicero, in his Philippics,[19] speaks sneeringly of Lucius Decidius Saxa, a Spaniard whom Julius Caesar made tribune of the people. The implication of his sneer is that Decidius started life as a skilled measurer of military camps, but had so risen above his station that he now hoped to measure out Rome itself as if it were a colony. We should not deduce from Cicero's words that there existed anything like the bureaucratic machinery with regard to surveying that we find under the Empire.

It was Julius Caesar who was in some ways looked back to as the founder of the profession. The late *Demonstratio artis geometricae* in the Corpus[20] says: 'Now let us come on to a letter of Julius Caesar's relating to the origin of this skill', and connects the expansion of surveying with Caesar's colonisation. Unfortunately the writer does not summarise the contents of the letter, and we have inadequate details. However, from the foundation charter of his Urso colony, mentioned above, one can see how careful was his attention to administrative detail in the foundation of colonies, which was always a prime concern of land surveyors.

The period following the civil war, when the Second Triumvirate had to provide land for thousands of discharged soldiers, was one of enormous expansion for surveying. One might even with some exaggeration say that this commission of three warlords, Antony, Octavian, and Lepidus, was a glorified land commission acting *ultra vires*. Certainly they were quite unscrupulous in seizing land wherever required, often acting in the name of Caesar, as having allegedly bequeathed them special powers. Some land was apportioned on a personal basis (*viritim*), but more often colonies were founded.

It was under Augustus that the first steps leading eventually to a large and elaborate bureaucracy were taken. Only from his principate onwards do we find organisation of centuriated land in a truly civil service manner. His name occurs many times in the Corpus. Much land was re-surveyed on his orders. Frontinus refers to a speech of his on the status of land belonging to *municipia*. Hyginus quotes an edict of his ruling that where land had been taken from a local authority and assigned to veterans, the colony's jurisdiction should extend only to areas actually made over to the veterans. Hyginus Gromaticus says Augustus ordered that in his allocations the numbers of *limites* should be inscribed on stones at all corners of 'centuries', and that he fixed the width of *limites*: for DM 40ft, for KM 20ft, for other *decumani* and *kardines* 12ft, for subsidiary roads 8ft. Colonies and boundary stones of his are frequently mentioned in the Corpus. He is said to have defined the terms 'excepted' (from local taxation) and 'granted', as applied to farms. Allocations of land under the Empire had to be made according to a law of Augustus, 'as far as scythe and plough have gone' (probably an old legal phrase revived by him), unless otherwise specified. The census of the whole empire, revised from time to time, included landed property; and the services of surveyors must have been needed to verify the accuracy of this.

The first *Liber Coloniarum*[21] says: 'To this list are to be added the measurements of *limites* and boundary stones from books not only of Augustus and Nero, but also of the surveyor Balbus, who in Augustus' times wrote up in notebooks the maps and measurements of cities in the provinces and who differentiated and published land law in its provincial variations.' Which book or books of Augustus are meant is not clear, but the breviary of the Empire and the description of Italy, two lost works by him, must have contained valuable material. No relevant work of Nero's is known, and one wonders if Tiberius or Claudius is really meant. As to Balbus, the only difficulty is that the Balbus who wrote the letter to Celsus which is preserved in the Corpus had served under Domitian or Trajan (p 42). Since there is no Balbus under Augustus who is known to have been actively concerned with surveying, it is most likely that the writer has made a mistake.

Gaius Julius Hyginus, who was Augustus' freedman and librarian, wrote a work, now lost, on the origins and topography of the towns of Italy. The extant works on surveying attributed to him must be later compositions. In fact one wonders if they were falsely attributed to him simply owing to an allusion to his work early in one of the treatises.

Augustus' principate witnessed a great blossoming of architecture, and the architectural manual in ten books by Vitruvius Pollio was written at this time. Vitruvius was a practising architect and insisted that the architect should be a man of the widest possible education, combining theory and practice, aptitude and sound instruction. By dignifying the architectural profession, he no doubt made it easier for the surveyor too to come to be regarded as a professional man of some standing. Much of Vitruvius' writing is inapplicable to surveying; but the description of the *chorobates* in 8.5, and the astronomy and orientation taught in Book 9, have a strong affinity with the theory and practice of land survey.

From the first century AD comes a tombstone of a Roman surveyor with a relief depicting his profession (p 50). The tombstone of Lucius Aebutius Faustus[22] is in the museum at Ivrea, North Italy (ancient Eporedia, said to have been founded as a colony in 100 BC). The marble slab has an inscription which may be translated: 'Lucius Aebutius Faustus, freedman of Lucius Aebutius, of the tribe Claudia, surveyor, *sevir*, erected this monument while still alive for himself and his wife Arria Aucta freedwoman of Quintus Arrius, and their children, and the freedwoman Zepyra.' At the top is a pediment with shield and spears, probably indicating that he had earlier been a military surveyor. Below the inscription is a dismantled *groma* or *stella* (p 69), whose staff is 73cm long and cross-arms 35cm. Above this are the symbols of a *sevir* (literally member of commission of six), one of a number of men selected for distinction by the local authority: two fasces with protruding axes, to denote Roman authority, and between them a settee with cushion and footstool.

The first century AD was a period when freedmen were continually bettering their position.[23] Of fourteen civilian land surveyors mentioned in inscriptions[24] eight were freedmen, three freedmen of the Emperor, one slave and two unspecified (presumably free citizens). Augustus' legislation created a class of *Latini Iuniani*, whereby slaves informally manumitted by their masters obtained statutory freedom; such a freedman could secure citizenship for himself and his wife when their child became one year old. Moreover Augustus and his immediate successors, especially Claudius, built up a new bureaucracy largely headed by freedmen. Although surveyors never attained the wealth and power of these imperial ministers, a freedman who was a skilled *mensor* of any type could under the Empire claim professional standing and might aspire to positions of some importance. *Mensores* came to be organised in guilds (*collegia*), which acted somewhat like their medieval counterparts. Rates of remuneration, which counted as a honorarium rather than salary or wages, would tend to become less haphazard, more related to those for teachers of mathematics and similar professional men. Rudorff explained this on historical grounds, in that land surveyors started as assistants to augurs, and so came to be classed with those who practised the 'liberal arts'. Although any surveyor was entitled to centuriate, it became illegal to allocate land without an appropriate qualification. If a surveyor or other measurer was alleged to have made a false measurement, the praetor could institute legal action against him.[25]

From the principate of Tiberius, generally associated with maintaining gains rather than expanding, comes the grandiose scheme for allocating huge areas of Africa Nova (southern Tunisia), mentioned on p 156. We should note that the centuriation was carried out by the third legion Augusta under the orders of the proconsul. This is an example where military surveyors put into effect a civilian scheme.

As has been seen, the reference to Nero as a writer on land topics is implausible: his talents seem to have run more to poetry. But quite possibly the Greek scientist Hero, or Heron, of Alexandria was writing at that time. Hero was known as the great mechanical engineer, who left his mark in many aspects of applied mathematics. His work *Metrica* deals with the measurement of regular polygons, circles, cones, pyramids, and other figures. He shows how to ensure accuracy in tunnelling a mountain simultaneously from both sides. His technical work which concerned surveyors is the *Dioptra*. The dioptra (p 76) was an instrument like a theodolite, capable of measuring vertical angles as well as horizontal, and thus of use not only to the architect or surveyor but to the astronomer. Hero claims not to be the inventor but the perfecter of a dioptra. His very elaborate machine does not seem to have been taken up by Roman surveyors. In addition to these published works, Hero made a name for himself as inventor of a perpetual motion device, penny-in-the-slot machines, mechanical aids to priests for imposing on credulous devotees, a fire engine, a water-organ, and so on.

The emperor Vespasian merits attention in an outline of Roman surveying. Frontinus[26] says of him: 'The emperor Vespasian actually raised money from some colonies whose *subseciva* had not been granted to them; for land allocated to no one could not belong to anyone except the person entitled to allocate it. By selling *subseciva* he made no small contribution to the treasury.' The standing of these remnants is given on p 94. In Chapter 11, where the Orange cadaster's are examined, it will be seen that in AD 77 Vespasian, so as to restore the State domains in the Orange area which had come to be occupied by private individuals, ordered a survey map to be set up. The inscribed stone tablets which resulted from this edict have greatly helped the study of centuriation and land revenue. Another inscription[27] records Vespasian as saying: 'I have now written to the procurator (of Corsica) telling him to define the boundaries and have sent him a surveyor.'

Sextus Julius Frontinus, whose writings on surveying have survived in extracts in the Corpus, was a very able man. Born about AD 30, he became urban praetor in AD 70 and after his consulship in 73 or 74 went to Britain as governor. As such he is best known for his successful expedition against the Silures of South-east Wales. His work on surveying was probably written, as were his two treatises on warfare, in the principate of Domitian (AD 81–96). Under Nerva, AD 97, he was appointed curator of the waters of Rome, and write a work on Rome's water supply which he completed under Trajan. This and the military works survive in full. Frontinus's knowledge of land survey was extensive, so that we may suspect he had been a land commissioner. It is only in the more technical aspects of the work on water-supply, such as the relation between diameter of pipe and capacity, that modern scholars have been able to call him a dilettante. Curiously enough, the emperor Domitian, who in the ancient historians comes in for little but abuse and in Martial and Statius for undiluted flattery, is considered in the Corpus a benefactor for having presented all *subseciva* in Italy to their occupiers. We also learn from Statius[28] of the preparatory work involved in Domitian's new road to Naples, which had embankments and deep cuttings: first the line of the road had to be marked out with furrows, then *limites* had to be cut (*rescindere*—'cut back'—*limites* is a technical phrase meaning to plot side turnings and cut a way through to them from the main road; hence simply to open a new road) and levelling carried out. It is this preparatory work, rather than the actual road-building, that must have preoccupied surveyors, though there is unfortunately no mention in the Corpus of this side of their work.

The Balbus whose treatise on shapes and measurements, in the form of a letter to Celsus, is preserved in the Corpus speaks of his recent participation under an emperor in a successful expedition against northern tribes. An adulatory phrase like *sacratissimi imperatoris nostri* cannot refer to Augustus and sounds far more like Domitian than Trajan. It was in AD 89 that Domitian, having personally conducted campaigns against two Germanic tribes, the Marcomanni and the Quadi, found it expedient to follow up a victory won over the Dacians two years earlier with an honourable peace recognising their king Decebalus. Scholars are apt to say that Balbus had been serving in Dacia, since he refers to that nation, but his words could perhaps better tally with participation in the expedition against the Germanic tribes. He is proud of the fact that the Romans would have been able to calculate the width of a river even if the enemy had been there to prevent them crossing.

Building surveyors at least were in great demand in Italy in Trajan's principate, AD 98–117. When the younger Pliny writes[29] to Trajan from Bithynia suggesting that money might be recovered from building contractors if an experienced *mensor* were sent out from Rome to help the architects, the emperor replies that there are not enough in Rome and neighbourhood to carry out the works which he is planning, but adds: 'But reliable ones can be found in every province, so you will not be without if you are willing to make a careful search.' The period was one of vast constructional enterprises, including baths, a new forum, the Basilica Ulpia, Trajan's Column with its wealth of detail on the Roman army, the harbour works at Ostia, Centum Cellae and Ancona, and probably the deep rock cutting for the new line of the Via Appia at Tarracina (Terracina). The treatise which comes down to us under the name of Hyginus, *De condicionibus agrorum*, was written in Trajan's principate. The writer says:[30] 'Recently a volunteer ex-legionary (*evocatus*), who had had sound military training and was most skilled in our profession, while allocating lands in Pannonia to veterans by the wish and generosity of the emperor Trajan Augustus Germanicus, adopted a new method on the bronze maps. He not only recorded the areas allocated, but at the end of the line relating to each settler gave the length, breadth and area of land allocated to each. This meant that no disputes about allocations of land could arise among the veterans.' Pannonia, south and west of the Danube between Vienna and Belgrade, was conquered by Tiberius in Augustus' principate and made into a province; under Trajan it was divided into Pannonia Superior and Pannonia Inferior. The colony concerned was almost certainly Poetovio (Ptuj, Yugoslavia), which is thought to have been founded between AD 103 and 106.

The *libri coloniarum* tell us that Trajan ordered boundary stones to be erected between Rome and Ostia in place of the old wooden boundary marks, and that he had a bronze map made of the area. So clearly he interested himself in civilian as well as military surveying. Although the two branches were separate, and although civilian surveying itself was divided into land and building spheres, all went hand in hand for many purposes. The younger men would tend to be enrolled as military surveyors, then acting on this experience turn to civilian surveying. And since the latter often involved settling ex-legionaries, who better than a volunteer ex-legionary to measure up and record their land?

Similarly in Hadrian's principate it was an ex-legionary volunteer who was called in by Quintus Gellius Sentius Augurinus, proconsul of Macedonia.[31] Hadrian wrote to him that he should settle a boundary dispute between the people of Lamia and the people of Hypata by calling in surveyors. He accordingly called in Julius Victor, an ex-legionary volunteer who was a surveyor. The inscription goes on to record the boundary then decided on as permanent. This record follows to some extent the pattern of a Delphi inscription in Greek and Latin,[32] which probably dates from the end of Trajan's principate. The composer of that inscription says that he himself was appointed judge by the emperor in a boundary dispute between Delphi and Anticyra. The line on which he decides, and which is specified in the inscription, is determined partly by his having found old boundary stones *in situ* and partly by a decision of Manius Acilius Glabrio as far back as 191 BC. This reversion to 300 years earlier, to the first time when the Romans had authority in Greece, is typical of the careful and conservative nature of Roman survey arbitration.

Hadrian is said by the *libri coloniarum* to have ordered boundary stones to be erected in the coastal area of Latium, numbered *terminus primus*, *terminus secundus* and so on. In AD 140 he legislated on the moving of boundary stones, distinguishing four different types of offence.[33]

With Hadrian, direct reference to the part played by emperors virtually ceases for a long period; and no writers in the Corpus after this period can be at all easily dated. Military surveyors are represented on the column of Marcus Aurelius in Rome as making arrangements for laying out a camp. In the third century AD there was a great revival of interest in Roman law, and we may perhaps ascribe the work of Siculus Flaccus, on types of land-holding, to that century. Outside the Corpus, but related to it, the treatise on camp-making is thought to belong to the same century. The importance of surveyors grew with their increasing employment as judges or arbitrators in cases of land dispute. Diocletian, emperor from AD 284 to 305, fixed the remuneration, among other categories, of each class of teacher: for every pupil, teachers of rhetoric received 250 copper *denarii* a month; teachers of literature, 200; teachers of geometry, also 200. This last category presumably included teachers of surveying, who however would have been specialised and different from the academic geometry teachers.

Three rulings by Constantine the Great, emperor AD 306–37, are preserved in the Corpus. There had evidently been cases where survey judges had abused their position. Thus one of these rulings is that a judge of a boundary dispute who is found to have claimed any of the disputed territory for himself shall not only lose the land he had no right to but an amount equal to this as well. Valentinian II ruled that in all cases of dispute over position of land the *agrimensor* (usual word for land surveyor in the late Empire) should be the final judge.

The later legal extracts in the Corpus include references, in edicts by Constantine, Theodosius, and Arcadius, to what was by then a very ancient procedure. Under the Lex Mamilia of the late Republic (p 104), land within 5ft of a boundary came under special consideration. This width was preserved the same down to the end of the Roman Empire. It applied in the East too: the last recorded extract referring to it is in the consulship of the emperor Arcadius and Rufinus: this was the despicable chief minister of the Constantinople court who was torn to pieces by his own troops.[34]

Under the late Empire, surveyors came to be organised more and more in the vast bureaucratic machine. Constantine the Great organised a whole staff of civil servant surveyors and other 'measurers' under a *primicerius* who was directly subordinate to the head of the *agentes in rebus* and indirectly under the *magister officiorum*. The term *primicerius* literally means 'first on the wax tablet', hence 'director', 'supervisor'.

After two years the *primicerius* tended to be promoted, as a knight, to the lowest grade of *agentes in rebus*. These last, of whom we first hear in AD 319, were imperial agents, organised in corps, who travelled round the provinces as couriers or (under Constantius II) virtually as secret agents, detecting evasion of taxes or treasonable actions.

Among honorific titles, *vir clarissimus* (not really implying that the recipient was 'very distinguished'), abbreviated to VC, was reserved for those with most meritorious service, while *togatus Augusti* (*Augustorum*) was a commoner distinction. Surveyors employed by the State were entitled to exemption from taxation under an edict of Constantine the Great. Each provincial governor had his own *mensores*; some of the more recently discovered evidence records those in North Africa. From now on we find Christian writers contributing to the Corpus, including a writer who speaks of Christ as one 'through whom peace between neighbours (*pax terminationis*, literally 'boundary-line peace') came upon earth'.

Justinian was particularly concerned with time limits on legal actions connected with land. Since by his time land titles might go back hundreds of years, it was essential that a maximum period should be fixed for each type of law-suit. Thus suits involving *limites* were restricted to a thirty-year limit, certain others to a twenty-year one. By his codification of the law, Justinian made it far easier for anyone to familiarise himself with the vast bulk of legislation, and this within a limited sphere was always the duty of a qualified surveyor.

Cassiodorus (ca AD 490–583) lived to a ripe old age, performed many important public duties, and by his writings serves as an important link between the ancient world and the Dark Ages. A letter of his,[35] dating from the reign of Theodoric, AD 493–526, shows that he knew more about land survey than some at least of his contemporaries. A boundary dispute, leading to violence, had arisen between two nobles. If, says Cassiodorus, they lived in a part of the Nile valley that was regularly flooded, they would simply call in a surveyor. Augustus's census of the whole empire, he continues, was intended to avoid such disputes (one imagines that was in reality a by-product); and the theory of surveying had been written up by Hero of Alexandria. Then Cassiodorus embarks on a pleasant academic sarcasm. Other sciences, he claims, are so theoretical that their professors address only a handful of students; 'but the *agrimensor* is entrusted with the adjudication of a boundary dispute that has arisen, so that there may be an end to wanton quarrelsomeness. He is a judge, at any rate of his own art; his law-court is deserted fields; you might think him crazy, seeing him walk along tortuous paths. If he is looking for evidence among rough woodland and thickets, he doesn't walk like you or me, he chooses his own way. He explains his statements, puts his learning to the proof, decides disputes by his own footsteps, and like a gigantic river takes areas of countryside from some and gives them to others'.

The last allusion to surveying in a writer on the fringe of the ancient world is in a letter of Pope Gregory I (Gregory the Great) dated July 597.[36]

> Gregory to John, Bishop of Syracuse.
>
> To prevent disputes about secular matters estranging the hearts of the faithful, great care must be taken that a dispute may be settled as easily as possible. We have learnt from Caesarius, Abbot of the Monastery of St Peter at Baiae, that a serious dispute has arisen between him and John, Abbot of the Monastery of St Lucia at Syracuse, about certain lands. To prevent this being prolonged, we have decided that it must be settled by a surveyor's ruling. We have therefore written to Fantinus the lawyer to send John the surveyor, who has set out from Rome for Palermo, to your brotherhood. We accordingly urge you to go with him to the areas in dispute, and by the presence of both parties on the spot to make an end to a dispute kept up by both sides despite [?] a limitation of forty years. Whatever is decided, your brotherhood should see that it scrupulously maintained, so that henceforth no dispute about the matter may reach us.

The letter goes on to say that Caesarius was inexperienced in secular cases and should therefore be given every help. Baiae, on the Gulf of Naples, was famous (and notorious) as a seaside resort in classical times, and, although remote, the lands claimed by the monastery must have been in Sicily. The word used for 'surveyor' is still *agrimensor*, and the letter shows that the profession was by no means dead with the fall of the Western Empire.

Thus, if we date the start of Roman land surveying from the start of Roman colonisation, while it is impossible to be precise, there are at least 900 years of development, reaching their culmination between about 133 BC, the tribunate of Tiberius Gracchus, and AD 138, the death of Hadrian. Their writings continued to be copied out by generations of monks, and the impact of their land division is still with us today.

4

The training of Roman land surveyors

GENERAL EDUCATION

The Roman educational system was heavily loaded on what we should call the arts side. As a form of training for the statesman, the writer, and many other occupations, it was well devised apart from certain defects; but it did not cater well for the technologist. In the initial stages, it is true, the pupil had to learn not only Greek and Latin, of which Quintilian actually thought Greek should come first, but also a fair amount of arithmetic. Horace's only complaint[1] is that in his local school in the colony of Venusia (Venosa) the arithmetic was too closely tied to money. The subdivisions of the monetary system were complicated and needed to be ingrained into the duller pupils. But after that poetry and rhetoric occupied the leading place, with Homer as the chief author studied. In the Silver Age the practice of rhetorical declamation, with *suasoriae* in which the student had to plead a course of action in some mythological or historical situation, and *controversiae* in which he had to argue on one side or the other, or both, of a highly fictitious legal case, so dominated secondary education that it left little time for the study of science and higher mathematics. If the student wished to round off his education, he would be sent to Athens, Rhodes, or Pergamon to study Greek philosophy or rhetoric, rather than to Alexandria where he might have learnt some applied mathematics. It is indeed a pity that the inventive genius of Hero was not coupled with the wealth, slave-power and business acumen of some of the Romans.

It only needed some such combination to invent an effective steam engine.

Nevertheless, the education available was far from narrow or wholly deficient in science. We can obtain the best idea of it by looking at a similar branch of learning, an account of the ideally educated architect by Vitruvius:[2]

> He must be both naturally talented and ready to learn. . . . He should be a literary man, skilled in painting, learned in geometry; he should know a good deal of history, should have diligently studied philosophy, have learnt music, not be ignorant of medicine, know the rulings of legal experts, and have a mastery of astronomy and cosmology. The reasons are as follows.
>
> An architect ought to know literature so as to give him a better background for writing up his notes. He ought to have a knowledge of painting so that he may give whatever coloured representation he wishes of his work. Geometry provides architecture with many resources: above all, it leads him on from rectilinear figures to the use of compasses, facilitating the planning of buildings on sites, with the accurate employment of set-squares, plumbline levels and straight lines. Through optics, lighting is correctly brought into buildings from specific areas of the sky. Arithmetic helps to calculate the cost of buildings, explain the workings of mensuration, and with the help of geometrical calculations and theorems to resolve difficult problems of symmetry. Architects ought to know plenty of history, because in their buildings they often include many ornaments, the reasons for whose existence they ought to be able to explain to enquirers. . . . Philosophy makes the architect high-minded and not arrogant but easy-going, fair, loyal and above all not greedy. . . . It also explains the principles of physics, necessary for many scientific problems, as for example in water supply. . . . An architect must know music so as to understand acoustics and the mathematics of harmony, and to be able to adjust catapults and other military engines. . . . There are also bronze acoustic vessels in theatres. . . . Also no one can make water-engines etc without a knowledge of music. He ought to know what is taught in medicine, so as to be able to take account of good or bad climate and geographical setting and the use of water. . . . He must know law, so as to be familiar with the regulations about ancient lights, water overflow and sewerage. Water ducts etc should be studied by the architect, who should take care before work is begun not to involve the owners in disputes, and to see that the rights of employer and contractor are safeguarded in contracts. By astronomy we learn the compass points, the order of the heavens, the equinoxes and

'e 49 (*above*) Fresco from Tomb of Menna at Sheikh Abd el Qurna, Thebes, Upper Egypt. In upper strip, surveyors with measuring equipment; in the lower strip, an official and scribes; ɔw *left*) Merkhet, Egyptian surveying instrument: holder, plumb-line and plummet; (*right*) suror's cross from the Fayyûm, a sighting instrument held in the hand, probably of the Ptolemaic period. The plumb-lines and plummets are restored

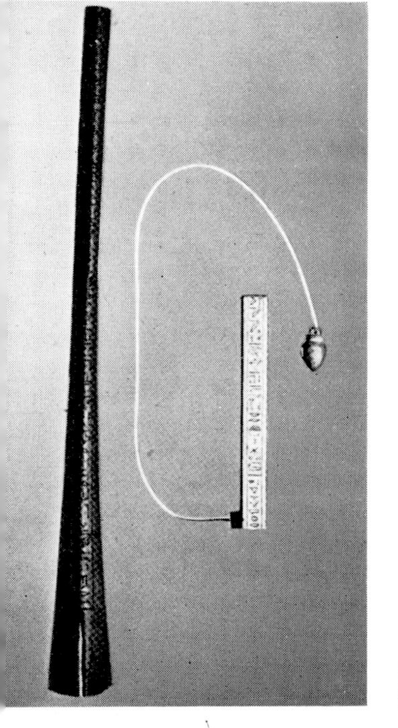

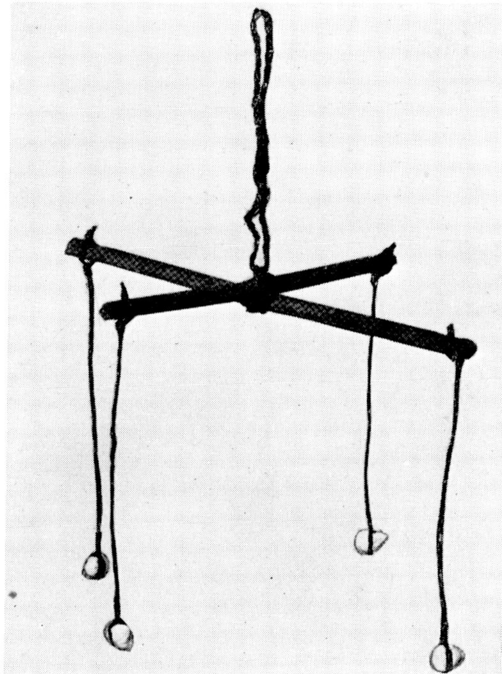

Page 50 (*left*) Tombstone of Lucius Aebutius Faustus, Ivrea, North Italy. The inscription shows that he and his wife were freedman and freedwoman, and that he was a surveyor and *sevir* (p 39). Below is a dismantled *groma* or *stella*, with the symbols of a *sevir* above it; (*right*) reconstruction of the *groma* whose metal parts, found in the workshop of the surveyor Verus at Pompeii, are now in the Museo Nazionale, Naples. The purpose of the bracket was to prevent the wooden staff obstructing sighting from one plumb-line to its opposite number

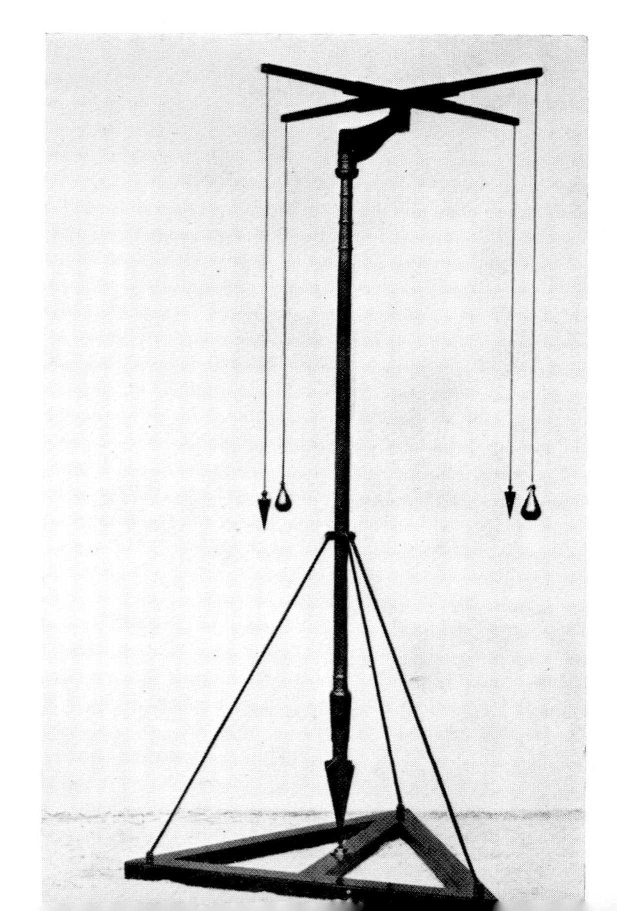

solstices, and the movements of stars. Without knowing astronomy he will not be able to understand how clocks and sundials work.

One may doubt if anything approaching such a good general education was either given to students of surveying or to anything like the same extent necessary for their work. Only in the sphere of Roman law was a regular and substantial body of teaching given in Italy which from the point of view of related subjects would have greatly helped the intending surveyor (p 63). In fact, as *agrimensores* came to arbitrate more and more on land questions, they needed to acquire a greater background of legal knowledge. The legal experts quoted in the Corpus are of various dates from the early to the late Empire, and cover quite a variety of topics.

The State took a gradually increasing part in organising education from the first century AD onwards. In the sphere of higher education professors of Greek and Latin rhetoric were appointed by Vespasian, and Marcus Aurelius set up professorships in Athens. Emperors also promoted school education: Trajan provided schooling for 5,000 children, and Antoninus Pius imposed on towns in the provinces the duty of providing schools. By the third century AD there was a fair measure of state control of education. Technical education was not included in the *artes liberales*, the subjects which were thought appropriate for a free man to learn, but since surveying was evidently, from its early connection with augury, regarded as one of the *artes liberales*, a senator like Frontinus could study it without being thought to descend to too humble an activity. Even so, it is amusing to find Hyginus Gromaticus[3] claiming a 'heavenly origin' for centuriation, or the Christian writer involving Christ in the preservation of boundary lines, but perhaps this at least had the effect of inducing generations of monks to copy out the Corpus.

GEOMETRY OF AREAS

The word 'geometry' literally means 'measuring of the earth', and to the writers in the Corpus it had precisely that connotation. A Roman surveyor had to learn mathematics of a practical nature to enable him to measure distances and areas, to orientate his surveys, and to make any calculations required by the government or local authorities for taxation or other purposes. But since surveying was regarded as one of the 'liberal arts', they were also given mathematical instruction of a more theoretical kind.

We are fortunate in possessing, in Book 5 of Columella's work on agriculture, an account of the type of mathematics used by a gentleman farmer in the first century AD. It has obviously inspired the treatise *De ingeribus metiundis* in the Corpus.[4] Lucius Junius Moderatus Columella came from a Gades (Cadiz) family which had long practised agriculture successfully in Baetica, made a province by Augustus. He served in a legion in Cilicia in AD 36 and afterwards acquired an estate which he farmed at Ardea in Latium. He gives[5] formulae for working out the areas of fields of various shapes, using not only *iugera* but subdivisions of them, which later fell out of use.[6] In fact even he ignores smaller subdivisions and starts with a half-*scripulum*, 50 sq ft (Roman); 576 of these make a *iugerum*, 28,800 sq ft. The chief complication in working out areas was not the somewhat clumsy Roman numerals (not quite as clumsy as some modern writers imagine: thus, 5,000 did not need to be written as MMMMM, it could be simply $\overline{\text{V}}$) but the fact that the Romans worked, like certain nations today, partly on a decimal system, partly on multiples of 8, 12, or 16. The *actus* of 120ft illustrates this. The complication comes out clearly in Columella's first example, a square field with sides of 100ft. Its area is 10,000 sq ft, but expressed in terms of *iugera* and their subdivisions it is a *triens* plus a *sextula*,* ie $\frac{1}{3} + \frac{1}{72}$ of a *iugerum* or 9,600 + 400 sq ft. The second example (Fig 3) is rectangular and presents no difficulties. It is a very standard

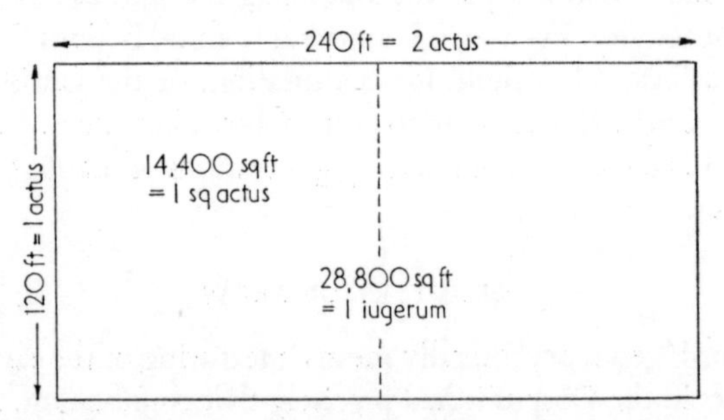

3. Measuring the area of a rectangular field, from Columella, *De re rustica*

* In the Loeb translation (Columella ii, 13), 'a *triens* plus a *sextula* of a *iugerum*' is not exact. A *sextula* is a sixth of an *uncia*, which in this sense is 2,400 sq ft. We should therefore translate 'a third of a *iugerum* plus a sixth of an *uncia*'.

measurement, since the sides are 2 and 1 *actus* respectively and the area is 1 *iugerum*. For a trapezoidal field (Fig 4) Columella averages out the unequal sides and gives 100 × 15 = 1,500 sq ft, which for a regular trapezium is correct.

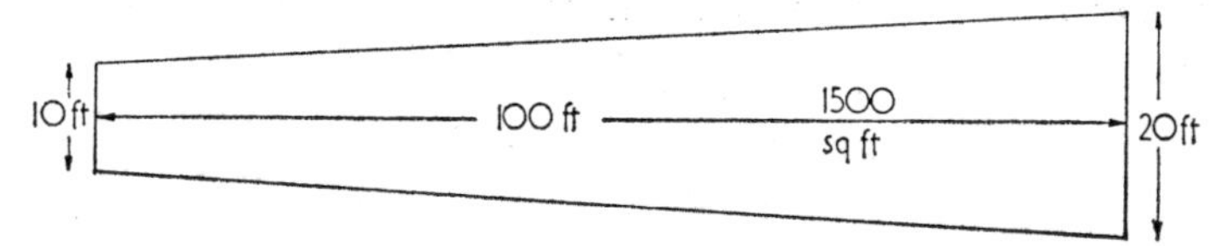

4. Measuring the area of a trapezoidal field, from Columella

Next he reckons the area of an equilateral triangle with sides of 300ft. He starts by squaring one side to make 90,000, and takes $\frac{1}{3} + \frac{1}{10}$ of this, 30,000 + 9,000 = 39,000 sq ft. In fact the correct answer is very close, $150^2\sqrt{3}$ = 38,971 sq ft. What he has done in effect is to take $\sqrt{3}$ as $1{\cdot}7\dot{3}$ instead of 1·73205. The other type of triangle he gives is a right-angled triangle with the sides adjacent to the right angle 50 and 100ft, area 2,500 sq ft.

For measuring a circle Columella uses, in effect, $\Pi = 3\frac{1}{7}$. His field has a diameter of 70ft, and he says: 'Multiply this number by itself: 70 × 70 = 4,900; multiply this by 11, making 53,900. I take one-fourteenth of this, 3,850, and say there are this number of square feet in the circle.' With a closer equivalent for Π, Πr^2 works out as 3848·46 sq ft. A similar method is used by him for the area of a semi-circle.

For the area of a segment (Fig 5) he says: 'I add the base to the height, making 20ft. This I multiply by 4, making 80, half of which is 40. Likewise half of 16ft, the base, is 8. The square of this is 64. I take one-fourteenth of this, making rather more than 4ft, and add it to 40; total 44, which I say is the number of square feet in the segment.' The correct answer is 44·730 to three places of decimals, while Columella's

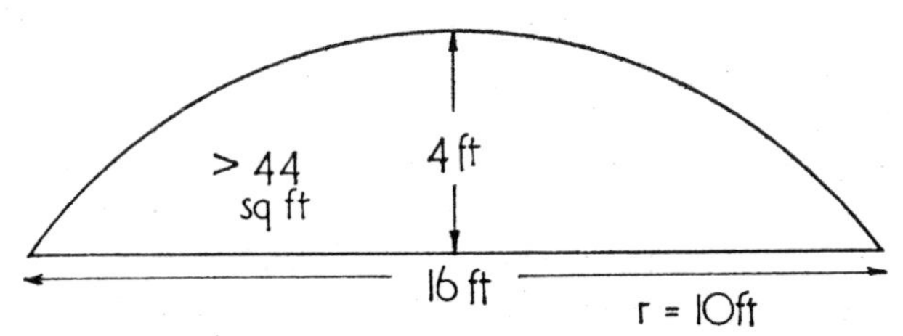

5. Measuring the area of a segmental field, from Columella

method actually produces 44·571. This is a very close approximation, but the method is hit-and-miss: it works exactly, assuming $\Pi = 3\frac{1}{7}$, only for a segment of which the corresponding sector constitutes one-quarter of a circle. With the correct value of Π, we can reckon the error to be as follows, with x = length, y = breadth:

x	y	*Correct value*	*Columella's value*
2	1	2·796	2·786
3	1	4·088	4·143
5	1	6·720	7·286
7	1	9·371	11·0

This shows that, as the length increases in proportion to the breadth, the error is magnified.

The same method, with different measurements, is found in the *De iugeribus metiundis*.

Clearly, if all land measurement had been carried out with such loose approximations as these, the land surveyors would not deserve to be praised for their accuracy. But whereas a segmental area corresponds roughly with the diagram of private land shown in the miniature illustrated on p 95(b), these are cases where accuracy was less important than with centuriated lands involving straight lines and right angles. An example of very accurate measurement is to be found in a section of frontier in Germany drawn in the principate of Domitian: for a section 29km long, the mean error of any point on the boundary is only about 2m[7]. Likewise the stretches of the Via Aemilia, in all 282km long from Ariminum (Rimini) to Placentia (Piacenza), and the centuriation orientated from it (p 146 and Figs 40, 41), were extremely carefully planned. If, therefore, the surveying treatise uncritically follows Columella in this error, it was no doubt because a segment was regarded as of minor importance for practical mensuration.

Whereas Hero in his *Metrica* explains how to measure a number of regular polygons, Columella contents himself with the above and a hexagon, each side of which measures 30ft. Although he does not say so, he clearly (contrary to the diagram in Forster and Heffner's translation and to that of the *De iugeribus metiundis*,[8] with the same measurements) intended it as a regular hexagon (Fig 6). 'I take the square of a side, 900; of this I take one-third, 300, and one-tenth, 90; total 390. This must be multiplied by six as there are six sides: the product is

2,340. So we shall say that this is the number of square feet.' This is based on the same method as for the equilateral triangle and is as close an approximation, the correct answer being $1350\sqrt{3} = 2338{\cdot}3$ to one place of decimals.

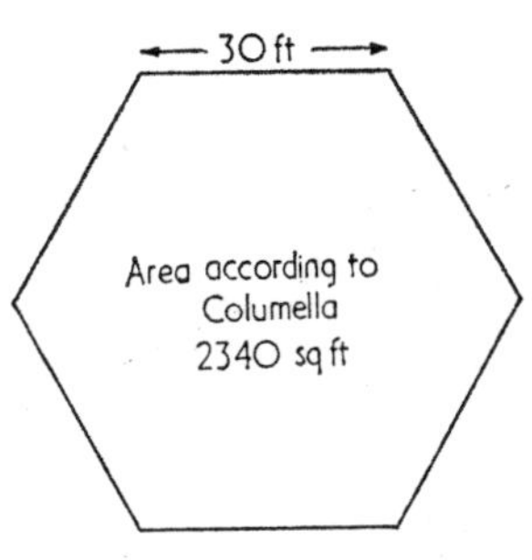

6. Measuring the area of a hexagonal field, from Columella

Whereas Columella does not directly use the theorem of Pythagoras, the Corpus does,[9] but some of its exercises are of an academic nature. Fig 7 is taken from Marcus Junius Nipsus, *Podismus*. 'Given a right-angled triangle whose perpendicular and base total 23[?ft], area 60, hypotenuse 17, find perpendicular and base separately. Solution: square the hypotenuse, 289; subtract 4 × area, 240, remainder 49; take square root, 7; add the sum of two sides, 23, total 30; halve, 15. This will be

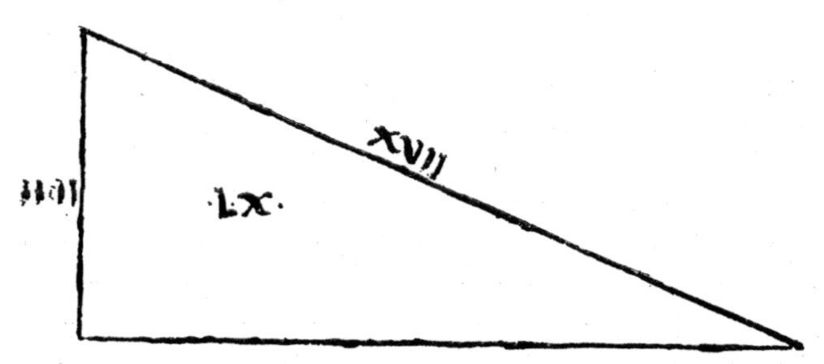

7. Finding the perpendicular and base of a right-angled triangular field, from Nipsus, *Podismus*

the base, so the perpendicular will be 8.' This is a highly academic problem, since it is most unlikely that one would have such data alone. The algebraic proof is as follows: a right-angled triangle with base a and perpendicular b will have area $\frac{1}{2}ab$; the square on the hypotenuse will be $a^2 + b^2$. From the latter subtract four times the former, giving $a^2 - 2ab + b^2$. The square root of this is $a - b$. Adding to this $a + b$ (given) we obtain $2a$ and so a. The ancients were slow in developing algebra, which was not, however, in itself an Arab invention; and only

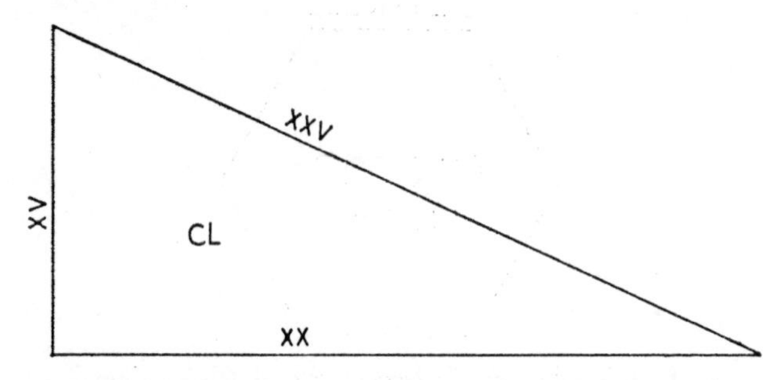

8. A similar exercise to that in Fig 7: given hypotenuse 25ft, area 150 sq ft, find the remaining sides. From Nipsus

in Diophantus of Alexandria (perhaps ca AD 250)[10] do we find an even moderately adequate algebraic notation. As a result, many examples later solved algebraically were given in arithmetical form with sample numbers, in this case the 'Pythagorean' basic sides 8, 15, and 17.

ORIENTATION

It was essential for the land surveyor to be able to find his bearings in the field. Even an augur, if he was scrupulous enough, would like to see that he really was facing the right direction.[11] Centuriation was sometimes orientated exactly or approximately from the compass points, sometimes from existing roads, sometimes (as in Dalmatia) from geographical features. Many *agrimensores* started as military surveyors, and for camp sites an orientation by the compass points was customary. The Corpus recommends it for all centuriation which starts away from a town. For this purpose only the minimum of equipment needed to be carried round. A portable sundial formed part of the equipment of the Pompeii surveyor, and something similar was obviously required by all of them.

Hyginus Gromaticus (p 130) is keen that the mid-point of the sun's appearance, not its rising or setting, shall form the basis of orientation.

'Many', he says, 'ignorant of cosmology, have followed the sun's rising and setting. . . . After positioning the *groma* under correct auspices, when the founder (of a colony) himself might be present, they have taken bearings from sunrise as accurately as possible. But then their *kardo* does not tally with the sixth hour.' The Roman system was to divide the daylight at any time of year into twelve equal hours. At midsummer in Italy these amounted to about $1\frac{1}{4}$hr each, at midwinter about $\frac{3}{4}$hr. The sixth hour, however, meant the end of the sixth, and this was always midday. Orientation from sunrise or sunset would also suffer if the land were hilly, so that the exact timing could not be checked (Fig 37).

The method recommended by Hyginus Gromaticus (Fig 9) is as follows: 'First we shall draw a circle on a flat space on the ground, and in its centre we shall place a sundial gnomon, whose shadow may at times fall inside the circle as well as outside. . . . When the shadow reaches the circle, we shall note the point on the circumference, and similarly when it leaves the circle. Having marked the two points on the circumference, we shall draw a straight line through them and bisect it. Through this we shall plot our *kardo*, and *decumani* at right

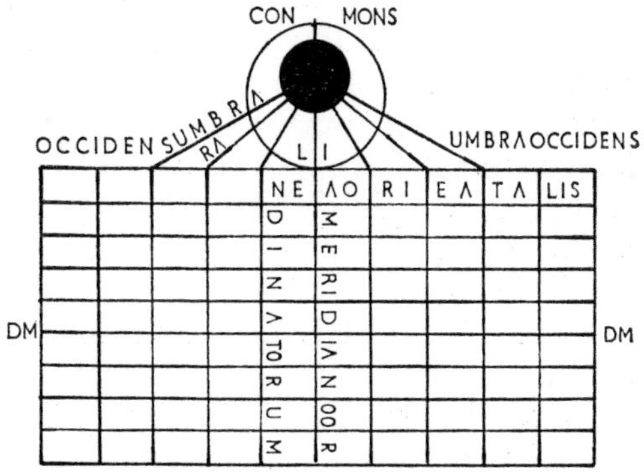

9. Hyginus Gromaticus. Orientation: a gnomon (corrupted to CON MONS), with east, south and west shadows. A line bisecting the east and west shadows is drawn towards the south and forms the *kardo maximus*. The *decumanus maximus* (DM) is then drawn at right angles to it. The diagram unsuccessfully tries to combine shadow ends with parallel north–south lines at equal intervals. On the right, *umbra occidens* is a mistake for *umbra oriens* ('east shadow')

angles to it.' This is the method also recommended by Vitruvius,[12] and it is still recommended for use by army units in the field.

An alternative method of obtaining South is mentioned by Hyginus Gromaticus. For this too a gnomon was required; three of its shadows were marked and measured. The solution, which depends on solid geometry, was first correctly explained by the map projectionist Mollweide.[13] A corrected three-dimensional plan was constructed by Dr C. Koeman.[14] The theory clearly goes back to Alexandrian scholarship. Perhaps the work of Apollonius of Perga (ca 262–190 BC) on conic sections, which is extant, led to some mathematical geographer thinking out this method.

SIGHTING AND LEVELLING

The procedure for sighting with the *groma* is described on p 70. The teachers of surveying were very strict on cross-checks to ensure accuracy. Such a cross-check, with sighting to rear and where possible sides, was scheduled at every *quintarius*, fifth intersection. (Since the Romans reckoned inclusively, it might be thought that *quintarius* should mean fourth intersection; but the diagrams show five *limites*, and a decimal system would be more helpful in working out the nomenclature of 'centuries'.) Hyginus Gromaticus says: 'The DM and KM must be plotted by very good surveyors, who must also fix the end of each *quintarius*, to prevent error which if accumulated will be difficult to correct. If there is a defect in the *groma* or the sighting, a wrong observation immediately appears on one *quintarius* and admits of fairly easy correction. Intermediate paths [*subruncivi*, literally ones which have to be weeded] are less likely to go wrong, but these too should be carefully plotted.' The diagram (p 68) purposely exaggerates the error in plotting. We do not, in the surviving centuriation areas, find major errors of planning.

A theoretical exercise in parallel lines is given by Hyginus Gromaticus (Fig 10). The text says: 'From base BD sight a landmark [G] and measure distances. Then FE:EG = FB:BA. Likewise sight on the other side and mark H at the point where CD exceeds the length of AB. Join HB. This will then be parallel to AC.' The object is to find out whether BD is parallel to AC and if not find a line running through B parallel to AC. For the purpose of the exercise it is evidently assumed that one cannot reach AC or measure AB or CD. The method therefore relies on similar triangles, whereas in practical surveying in antiquity triangles were little used. The diagram distorts by placing H near C, whereas it should be near D.

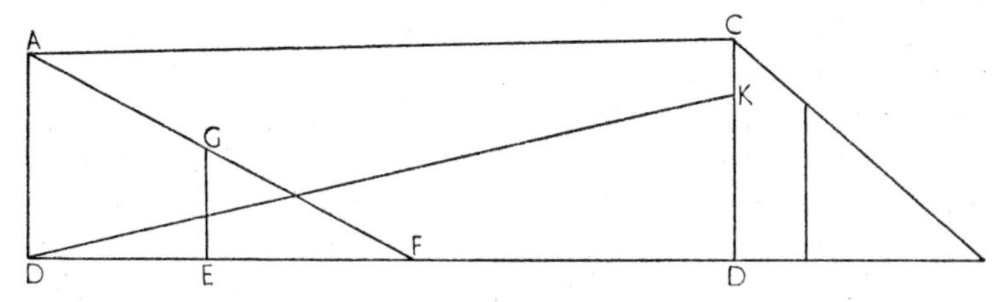

10. Theoretical exercise in parallel lines, from Hyginus Gromaticus (the scribe has two errors: D on the left for B, and K for H)

There is mention in the Corpus of levelling, indicated by the technical verb *cultellare*, literally 'cut with a knife'.[15] Frontinus[16] says: 'If there is a valley which exceeds the surveyor's angle of vision, he will have go down through it by putting marks near the instrument, ie *groma*. So as to make the approach to this line equal to its opposite number, we shall have to draw a horizontal line to the highest point of the survey, then drop a perpendicular to the measured line. Whenever we measure without a line, losing sight of the marks and then catching sight of them again as we hurry from the spot, there is as much wastage in that change of survey, however slight, as if we were following ground level. But if there is a smaller valley, too far to sight across, in order to avoid difficulty we may walk across it, call out not less than three marks on the other side, then checking these turn the instrument round and look back at the earlier marks, and by suspension carry on the line we have begun as far as required.' The phrase 'call out marks' (*dictare metas*), which occurs elsewhere too, implies that when the assistant was fixing a pole at the right distance on the sighting line he called out to his companion at the *groma*, in case any correction was needed. By measuring the horizontal lines the width of the valley could easily be calculated. Alternatively, one presumes, the less accurate means used for calculating the width of a river could be adapted to a valley—see below.

DISTANCE CALCULATION

A short work by Marcus Junius Nipsus, *Fluminis varatio*,[17] is concerned with working out the width of a river. The method could be used for any calculation of distance. We do not know the date of Nipsus (his Latinity suggests a late date), but the principle was familiar to the surveyor Balbus in the northern campaign of which he was on the staff. The text of Nipsus on this goes: 'If as you are squaring land you have a river in the way which needs measuring, do thus. From a line perpendicular to the river make a right angle, placing a crossroad sign. Move the *groma* along the line at right angles, and make a turn to the right at right angles. Bisect the line from this right angle to the original one, and place a pole at the mid point. With the *groma* here, make a line in the opposite direction from the pole you had placed on the other side of the river. Where this meets the line drawn at right angles, place a pole, and measure the distance from here to the crossroad sign. Since you have two triangles with equal perpendiculars, their bases too will be equal. So you now know the distance from the pole across the river to the first crossroad sign; subtract the distance from the latter to the river, and you have the width of the river.'

The manuscript diagram is shown in Fig 11, but this does not tally

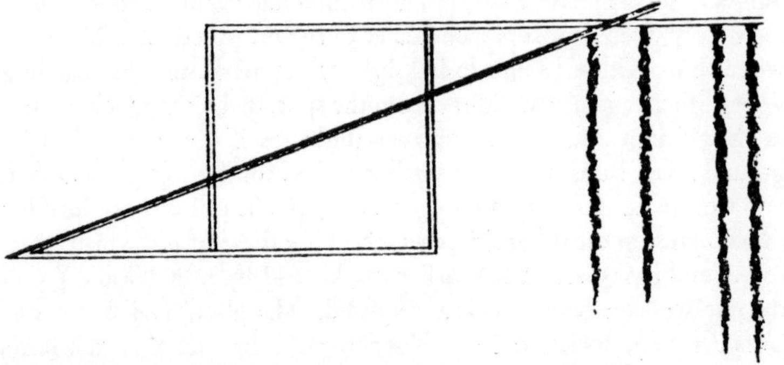

11. Measuring the width of a river (given as wavy lines on the right). The diagram is inexact, and has been emended as shown in Fig 12. From Nipsus, *Podismus*

with the text. There is no bisection on it, and in order to make the scheme work, one of the triangles should lie athwart the river. Hence Lange[18] was right in re-drawing it as in Fig 12.

Nipsus also writes of *varatio* in centuriated land. Unfortunately the text is so corrupt and so different from the diagram (Fig 13) that no sense can be made of it. But it looks as though he is recommending a cross-check by sighting on a diagonal at repeated intervals.

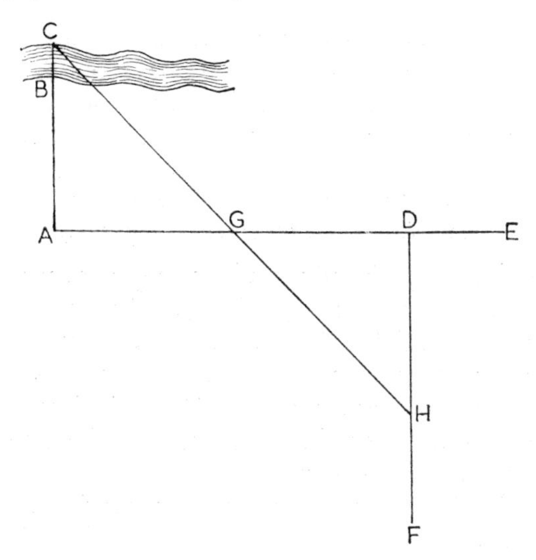

12. Reconstruction of diagram intended by Nipsus (see Fig 11) by C. C. L. Lange. CAG, GDH are congruent triangles

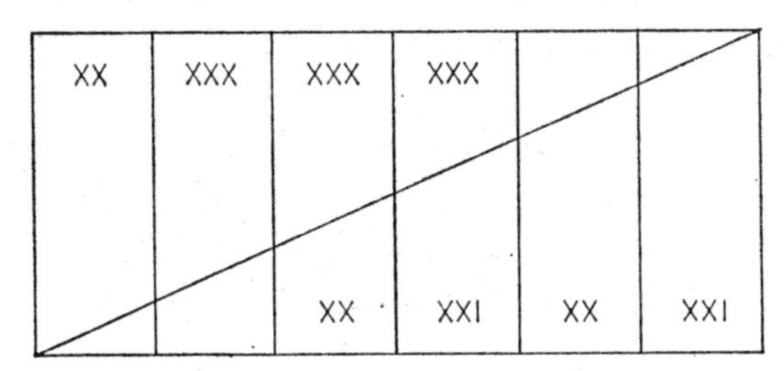

13. Cross-check on survey by sighting on a diagonal, from Nipsus

ASTRONOMY AND COSMOLOGY

The training of land surveyors included a fair amount of teaching about the earth and the universe; some of this has come down in the Corpus. The predilection of the ancients for using poetry as an educational medium comes out here too, since Virgil[19] is quoted on the five zones and Lucan on the southern hemisphere. The origin of the five zones theory was Eratosthenes in his poem *Hermes*. The names of them in the Corpus, from north to south, are *septentrionalis*, *solistitialis*, *aequinoctialis*, *brumalis*, and *austrinalis* (northern, solsticial, equinoctial, brumal, and southern). Only the temperate zones, the second and fourth, are thought to be properly habitable. The fourth is inhabited by Arabs, Indians and other races; as Lucan[20] puts it of participants in the civil war:

> You, Arabs, came into a sphere unknown,
> Surprised to find the shadows on your right

If they looked west in their native southern temperate zone, they would have the midday shadows on their left; in fact they are looking west as they march through Asia Minor to Greece, so have north on their right. From these quotations it is clear that the teachers of surveying thought in terms of a spherical earth; and Vitruvius is even, following Archimedes, prepared to say that the curvature of the earth may affect the water surface in a water-level.

On the other hand they thought of a geocentric earth: Hyginus Gromaticus gives us a Pythagorean version of this (Fig 14). 'The cosmologists', he says, 'describe the earth as a point in the sky by saying that from the pole to Saturn's orbit is the interval which the Greeks call a semitone; from Saturn to Jupiter a semitone; from there to Mars a tone; from Mars to the sun, three times the distance from the pole to

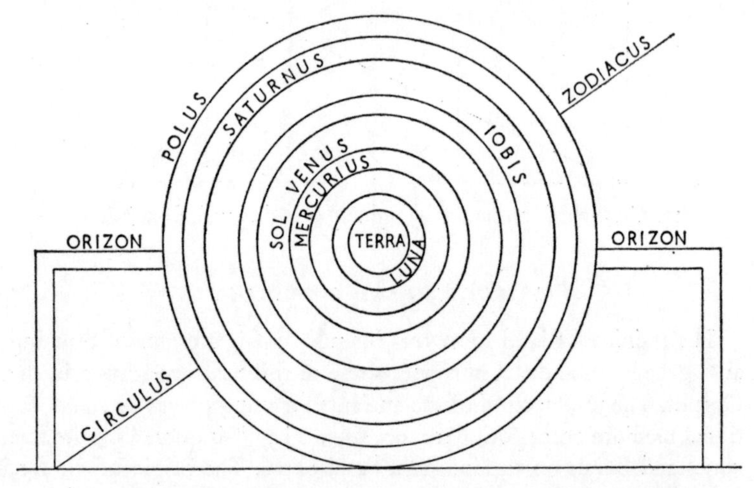

14. Orbits, round the earth, of the moon, Mercury, Venus, sun, Jupiter, and Saturn, and outer limit of heaven: Hyginus Gromaticus

Saturn, ie $1\frac{1}{2}$ tones; then from the sun to Venus, as much as from Saturn to Jupiter, a semitone; then from Venus to Mercury a semitone; then from Mercury to the moon the same amount, a semitone; from the moon to the earth, as much as from the pole to Jupiter, a tone.' Pythagoras, having discovered musical harmony, tried to apply it to a complete harmony of the spheres, and the neo-Pythagoreans had many followers in Rome.[21]

THE LAW AND THE SURVEYOR

Of the many spheres of Roman law, the surveyor had to be familiar with only two in great detail. One was the body of law governing the classification of land, the other was that concerning boundaries and boundary disputes. Other aspects, such as the law of inheritance or the procedure for obtaining restitution, might well concern them peripherally. As professional men they would want to avoid implication in actions under the law of fraud, including false measure. In case of threats to public roads or rivers, they might well have to apply for an interdict. But the two first mentioned were their prime concern.

Siculus Flaccus begins his treatise: 'It is well known, even to men who have no connection with our profession, that the status of land is different all over Italy; and we frequently find this in the provinces too.' He rightly explains differences in the status of local communities on historical grounds: some have bitterly opposed Rome, while others have been her allies from the start. Some places were colonies, some *municipia*, some *praefecturae* (originally assize-towns, mostly assimilated to *municipia*). The status both of countries and of towns throughout the Roman Empire had to be understood at least in principle.

As to land,[22] the surveyors classified it both in a strictly professional way and in the terms in which it was described in land laws, etc. The professional classification was as follows:

1 Parcelled and allocated land, divisible into:
  (a) centuriated land;
  (b) land allocated in *scamna* and *strigae* (p 94).
2 Land measured by its boundaries.
3 Unmeasured land.

The details of 1(a) and (b) have been given in Chapter 6; they occupied pride of place in the surveyor's list. But he had to recognise that this classification was chiefly for his own convenience. The law-courts would above all need to know the status of the land and of buildings on it. For example, was the land *ager publicus*? If so, in the territory of what community did it lie? Was it centuriated, and if so since when? Was it subject to taxation? Were there any farms exempted from taxation? Was there a legal limit to holdings, and if so had any exceptions to this been made?

If we are to go by Frontinus's definition (p 95), holdings of *strigae* and *scamna* were allocated on a system of straight lines without width (*rigores*) separating them from neighbouring plots. As Weber[23] saw, the land register would be likely to contain lengths and breadths of all such holdings; whereas in the case of centuriation, wherever the holdings within a 'century' were identical in size, only their areas, not their linear measurements, would normally be recorded: the case of a colony in Pannonia (p 42) was exceptional.

Legal definitions were particularly important in law-suits affecting property. In Fig 15, for example, the central area is not merely marked

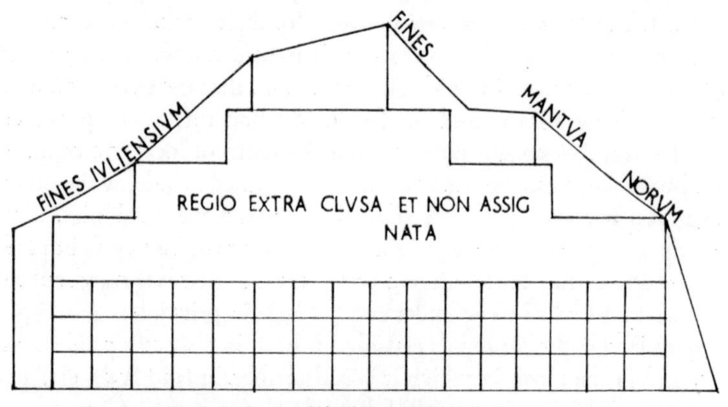

15. Plan inscribed with the legal definition of the status of an area: Hyginus Gromaticus

extra clusa but *extra clusa et non assignata*, 'excluded from centuriation and not allocated', to make the position quite clear. At the same time the surveyor had to be taught to apply reason and make allowance for custom. Thus Frontinus, on disputes about area (p 106, no 5),[24] says: 'When, in allocated land, area is being considered in relation to the map, it is customary for the date of allocation and the cultivation to be looked into. If memory of the change (of ownership or status) has already faded, the law's formula usually intervenes and prevents surveyors from entering upon such disputes. . . .' Again, Siculus Flaccus,[25] speaking of boundary marks, says: 'Particular attention should be paid to regional customs, and precedents should be drawn from neighbours.' As an example of regional differences, Frontinus[26] says: 'Many different incidents relating to ordinary law arise because of the diversity of provinces. In Italy a pretty big dispute may flare up in order to keep off rain water. But in Africa the same issue is handled quite differently. Since that is a very dry area, they have no dispute on this score unless someone has stopped rain water flowing on to their land. They make embankments, and catch and retain the rain water, so that it may be used up on the spot rather than flow away.'

We also find regional differences stressed in differentiating types of boundary. On this, Siculus Flaccus says:[27] 'Sometimes one can gather from the actual appearance: if someone builds walls on his own lands to keep up and preserve his fields, they cannot be boundary walls. Some walls which do show boundaries are found to be built more solidly than ones built for private purposes. Nevertheless, in this type of definition too you will have to look at regional customs. But sometimes one can deduce a certain amount from the nature of the areas themselves. If there are no lands for which a wall seems to have been made as support, it may be thought to be a boundary wall. But in flat areas if a field is stony it is cleared, and from the heap of stones walls are made.'

The aim in this chapter has been to consider the educational impact of law, as of other subjects, on men being trained as Roman surveyors. The detailed examination of Roman land law is a specialised topic. The fact that there were so many different types of holding and so many types of land dispute, often governed by laws of very varying dates, meant that much legal detail had to be memorised and much excerpted for reference. The codex form of book made this much easier than the earlier papyrus roll.

5

Roman surveying instruments

'GROMA' AND 'STELLA'

The principal Roman surveying instrument was called *groma*. The word is derived from Greek *gnoma*, akin to *gnomon*, almost certainly by way of Etruscan: the change of consonant can be paralleled. There was also a shift of meaning, since *gnomon*, as far as surveying was concerned, meant a sundial pointer, whereas *groma* meant something like a cross-staff. It was regarded as the instrument most typical of the surveyor, so that it appears in stylised form on the tomb of Lucius Aebutius Faustus (p 39). It was not until late Latin that surveyors themselves were called *gromatici* after it. The instrument was used in military as well as civilian surveying, and we are told that a central point in a military camp was called *gromae locus*. Since this was a point on the ground (and presumably also on a plan) where a surveying instrument was used, it may be compared to the modern trig point. It will be seen below that with the Pompeii *groma* the staff was offset slightly from the exact centre. The writers in the Corpus sometimes speak more vaguely of this or a similar instrument as *ferramentum*, 'iron instrument'.

Since the Corpus does not give a picture of the *groma*, earlier scholars did not have a very clear idea of its exact appearance. The tombstone mentioned above, that of Lucius Aebutius Faustus of Eporedia (Ivrea, north Italy) has a relief showing the instruments of his profession (p 50). The staff of the surveying instrument is upright; the cross is detached, perhaps to symbolise the end of his life-work, and laid diagonally over it. Was the surveying instrument on this monument a *groma*? Since the representation on the tombstone is only schematic, one cannot be sure.

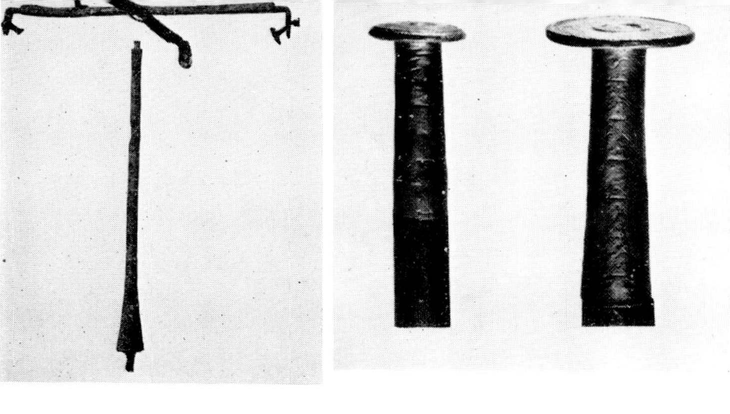

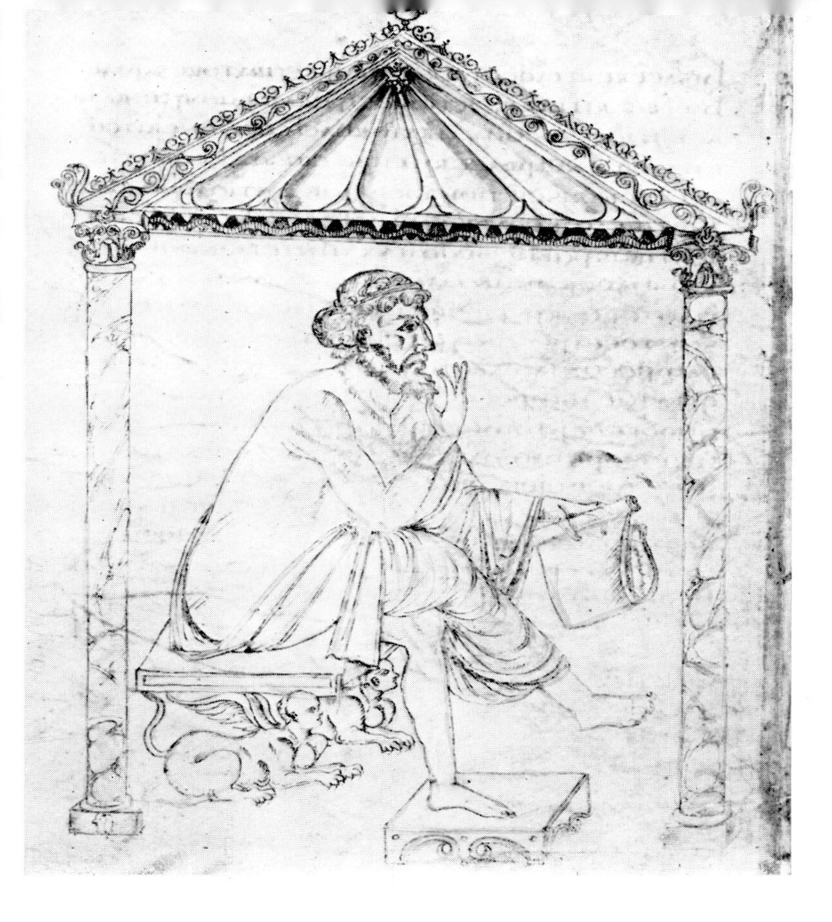

Page 67 (*above left*) Roman surveying instrument found at Pfünz, near Eichstätt, Bavaria. This was thought by Schöne to have had a wooden frame round the cross and a wooden base; (*above right*) iron end-pieces of a wooden measuring rod, *decempeda* (10ft long) or *pertica*, found at Enns, North Austria; (*right*) seated geometer or surveyor with scroll. From MS A, Wolfenbüttel

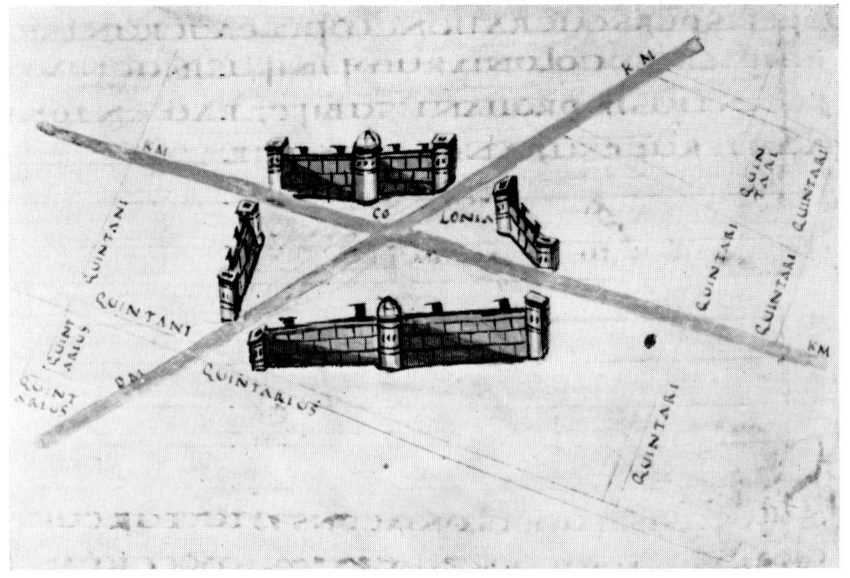

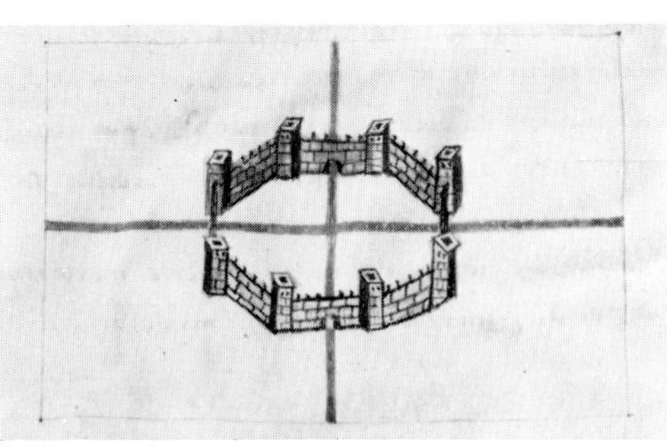

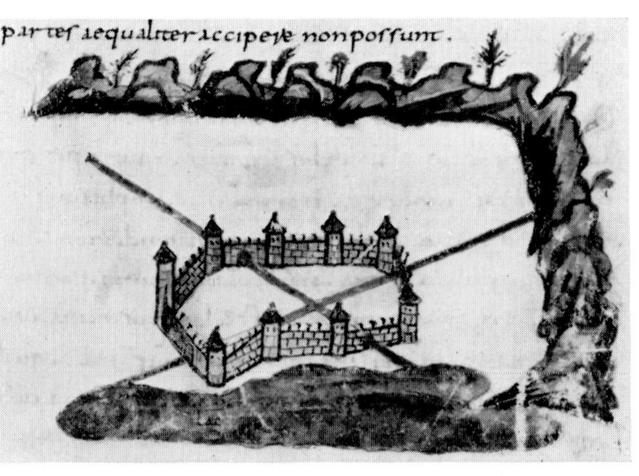

Page 68 (*above*) What happen
when the *kardo maximus* (KM) an
decumanus maximus (DM) at on
end wrongly marked KM, are no
laid out at right angles owing t
faulty survey. The *quintarii*, ever
fifth road from the centre, wi
show up the defect. (Hyginu
Gromaticus) from MS A, Wolfenbüttel; (*middle*) Admedera (Hai-
dra, Tunisia). The centre of th
survey is, exceptionally, in th
exact centre of the colony
Miniature (Hyginus Gromaticus
from MS P, Vatican; (*below*)
colonia by the sea, which cuts o
the KM and DM in one directio
each. Miniature (Hyginus Gromaticus) from MS P, Vatican

There are in effect two names for three stages of complication. The names are (a) Greek *asteriskos* = Latin *stella*, (b) *groma*, Greek *gnoma* (Suda). The stages are (a) the wooden cross found in the Fayyûm (p 27), (b) the iron cross found at Pfünz, (c) the Pompeii *groma*.

The metal portions of a surveying instrument were discovered at Pfünz, near Eichstätt, Bavaria.[1] They consist of a rigid iron cross with a hole at the intersection. Each of the four arms has its end bent down and a strong iron nail inserted in a hole. The instrument is thought to have been operated with a wooden frame round the cross and a wooden base of some kind. Schöne reconstructed it with a square wooden frame, 35 × 35cm, consisting of four wooden slats at right angles to the iron arms and dovetailed together. The iron shaft is only 35·5cm long, so needed a base, but perhaps not, as Schöne thought, a tripod base: it may well have had a pointed staff as base. The disadvantage of an instrument of this type is that the staff obstructs accurate sighting from one plumb-line to its opposite number.

The only example of what we imagine to be a true *groma* is the one whose metal parts were found in 1912, when during the excavations at Pompeii the workshop of a surveyor called Verus was discovered. We do not know his other names; there is no reason why he should, as Della Corte thought, have been the same as a Verus recorded in a Pompeii electoral scribble. The main find[2] was the metal parts of a *groma*: the wood had perished. The original finds are in the National Museum, Naples. The *groma* was reconstructed by Della Corte (p 50); a copy of his reconstruction may be seen in the Science Museum in London, and simpler wooden models have been made for the University of Glasgow and the National Museum of Wales at Cardiff.

At the top is the cross, which has iron sheeting; this originally enclosed wooden arms. To prevent inaccuracy due to the wearing away of the wood, the arms were reinforced near the centre by bronze angle-brackets. Through a hole near the end of each arm hung a plumb-line. The four plumb-bobs which hung at the ends of these are not identical in shape, but are in two pairs, obviously arranged at opposite corners, as shown on p 16. Examples of plumb-bobs may be found in a number of museums, but they do not necessarily come from a *groma*, and where finds of them are isolated, they are most likely to come from a mason's plumb-line. Such a find was made recently at the training excavation at the Roman camp near Bainbridge, Yorkshire.

The system of sighting from one plummet to its opposite number worked most effectively if the cross was off-centred; otherwise, as with the Pfünz instrument, there would be an obstruction. To avoid this, the cross was placed not directly on the staff but on a bracket. The parts of the bracket that are preserved are the curved iron supports from its top and bottom; the core was made of wood. The bottom of the bracket fitted into a bronze collar set into the top of a wooden staff. The horizontal distance of the centre of the cross from the staff was calculated by Della Corte as 23·5cm; one might have expected it to be 1 Roman foot (29·57cm). The length of the staff is not known, but may have been about 2m. A fluted iron shoe was attached to its base.

The method of operation was for the surveyor to plant the *groma* in the ground the bracket length away from the required centre of survey. He then turned it until it faced the required direction, which he had ascertained beforehand, finding south either by the methods described on pages 86–7 or from a sundial. Sighting was done by looking from one plummet to its opposite number, the plumb-bobs being grouped in two pairs to avoid confusion. Sights could be set on to a second *groma*, positioned first perhaps 1 *actus* (35·484m) away, then a similar distance from the first and second *gromae* at right angles. The square would then be completed and cross-checks taken.

The *groma* had only a limited use: it enabled straight lines, squares and rectangles to be surveyed. But these were exactly what the *agrimensor* normally required, so that more complicated equipment was unnecessary on a straightforward survey. If there was not too much wind, the *groma* would work adequately for this purpose; and no doubt a temporary wind-break could always be erected. If the cross-check caused doubts, surveying could be postponed until the wind dropped. But this deficiency was spotted by Hero of Alexandria. He says that the *asteriskos* ('starlet') may prove inadequate in a strong wind; which of the three types he is thinking of (the Pompeii *groma* as well as the other two types could be called a starlet) is uncertain, but it may have been the simple Fayyûm type.

PORTABLE SUNDIAL

Among other instruments found in Verus' workshop was a portable sundial. The Greeks experimented with both flat and concave sundials, and the Romans relied on Greek inventions. The ancient system was to divide the daylight hours into twelve throughout the year. The shadows cast by the gnomon vary according to the latitude and according to the season. Yet we are told by the elder Pliny[3] that the first sundial to reach Rome was one captured in 263 BC at Catana (Catania, Sicily), and for nearly a century the Romans used it in Rome without realising that it was designed for a more southerly latitude. Then in 164 BC a new model corrected for the latitude of Rome was constructed. The most helpful of the flat type were those resembling the Delos sundial which catered for each hour at midsummer, the equinoxes, and midwinter. Vitruvius[4] reckons that at Rome the midday shadow at the equinoxes is eight-ninths the length of a gnomon, and gives instructions for constructing a sundial suitable for the latitude of Rome.

Roman portable sundials, many of which lacked such refinements, came in various shapes. The Pompeii model (Fig 16) is in the form of an ivory box, 10·6 × 5·2 × 3·8cm, with a small silver clasp. The portable gnomon which was used with it was not discovered. The box

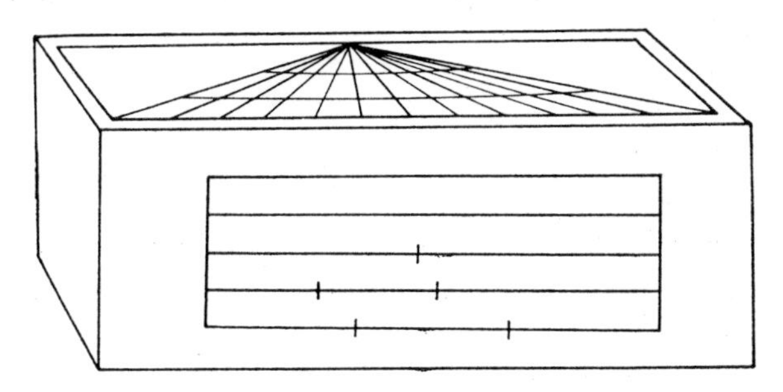

16. Ivory box from Pompeii, 10·6 × 5·2 × 3·8cm, serving as portable sundial and measure

is marked off on the upper surface with thirteen converging lines, which showed the daylight hours from first to twelfth without seasonal correction. These converging lines are intersected by two straight lines. A somewhat similar type of portable sundial, but circular and with limited provision for latitude and season, is in the museum at Aquileia.[5] It has monthly lines on both sides, one marked RO for Rome, the other RA for Ravenna. One such sundial (Fig 17) was found at Crêt-Chatelard in the Department of the Loire, with built-in gnomon.[6]

An extremely similar one is preserved in the Oxford Museum of the History of Science.[7] On this the latitudes given for various provinces are mostly a mean between the extremes given by Ptolemy. That allotted to Britain is 57°, which brings us nearly as far north as Aberdeen. It is thought to date from the third century AD. Roman soldiers

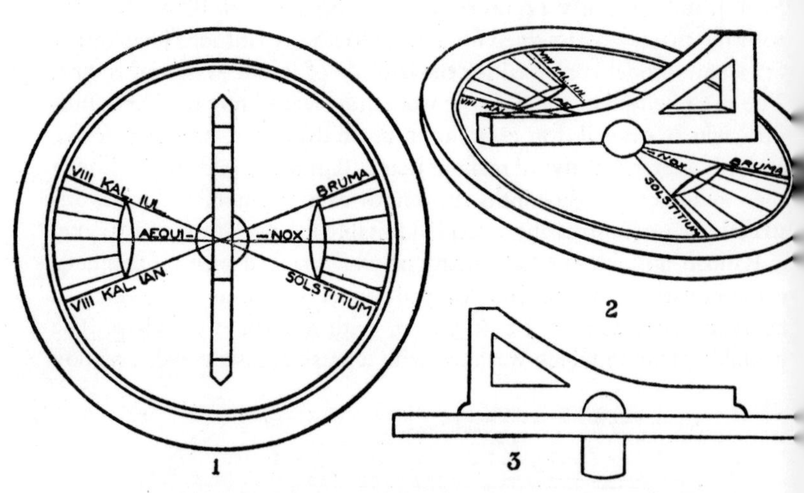

17. Portable sundial from Crêt-Chatelard (1) from above, (2) oblique view, (3) from side

reached nearly that latitude in the Severan campaigns; even so there is exaggeration. It is thought by Stebbins to take account of Roman unequal hours, which may be true, and by Price[8] to be self-orientating, which is disputed. This flat type, however, placed horizontally with a gnomon, is perhaps not likely to have been the commonest. In addition to its one specimen of this type, the Aquileia museum has no fewer than six concave Roman sundials, some more adorned than others. These consist of a concave hemisphere on which divergent lines are traced. The sundial was held vertically, and the sun filtering through a small opening in the top showed the time on the divergent lines. A discussion of the various types of Greek and Roman portable sundial is to be found in Price,[8] who, however, omits the ivory box from Pompeii and mentions only one Aquileia model.

The Pompeii box also served as a measure: two of its sides have lines with marks on them indicating measurements. On one side these are indistinct; on the other, line 4 has two marks, each designating a digit, 1·8cm; line 5 has three marks, each for a Roman inch, 2·46cm. What the mark on line 4, 3·4cm from the end, stands for is uncertain.

MISCELLANEOUS EQUIPMENT

Other finds in the Pompeii workshop were: (a) a folding ruler with one fold, 1 Roman foot long; there are other examples in the Naples museum; (b) two bronze compasses, one with iron points; lengths 12·8 and 13·3cm; (c) the bronze parts of two wooden chests, one elliptical and one cylindrical; a key was found with them, which Verus no doubt used when he did not want the general public to see or pilfer his notebooks, etc; (d) pincers, 9cm long; (e) a conical ferrule, 6·2cm long, of uncertain use; (f) a bronze cone, 14cm long, open at the side and with a small ring soldered on its edge; the use of this too is uncertain; (g) sixteen long, thin iron tools. By way of writing equipment, an ink bottle and a stylus were discovered. These suggest that Verus used wax tablets for notes, calculations and rough drafts, whereas he used papyrus rolls for reports and for the type of finished drawing that did not have to be engraved on bronze.

End-pieces of a measuring-rod were also found at Pompeii, but better preserved ones were discovered at Enns in north Austria. These belonged to a wooden measuring-rod, *decempeda* or *pertica*, the first name indicating a length of 10 Roman feet. The Enns end-pieces[9] were three in number, two belonging to one rod and one to another. The first two are each 8·5cm long, somewhat tapered, and with flat circular ends, which enabled the rod to be aligned with another of the same dimensions. Thus if an *actus*, 120ft, needed to be measured, two 10ft rods each used six times would suffice. To help the surveyor with smaller measurements, the first two of the end-pieces were marked off in inches (27·75 or 27·9mm) of the *pes Drusianus* and in half inches (12·45mm) of the *pes monetalis*. The other end-piece was marked off in inches of the same type and in digits or finger-breadths, sixteen to a foot, of the *pes monetalis*.

Hero says that ropes or cords, twice tautened, then smeared with wax, were used for measuring long distances. None has survived, and in pictorial representation they are seen only on frescoes of dynastic Egypt.

'CHOROBATES'

The *chorobates* (end and side views Fig 18) was used mostly for the levelling of aqueducts, and is in fact never mentioned by the writers in

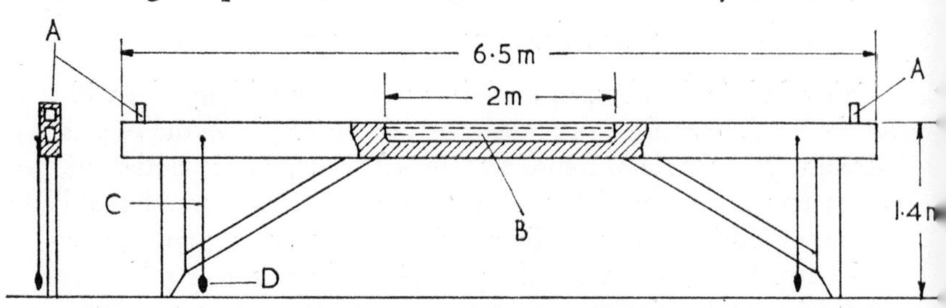

18. Reconstruction of *chorobates*, levelling instrument (not from Vitruvius' measurements). No examples have survived. A: sights; B: water channel; C: plumb line; D: plummet

the Corpus, although they speak of levelling (p 59). Vitruvius,[10] explaining the supply of water to country houses and cities, says:

> The first stage is levelling. This is done by dioptras or water-levels or by the *chorobates*. This last is the most accurate, since dioptras and levels are liable to error. The *chorobates* is a straight rule about 20ft long. At the very ends it has equal legs fastened to the rule at right angles, and between the rule and the legs there are cross-struts fixed by tenons. These have perpendicular lines accurately drawn, and plummets hanging from the rule, one over each cross-strut. When the rule is in position, the perpendiculars, touching the marks equally, show the position as level. If the wind obstructs this and if because they move about the lines are unable to give a definite result, the *chorobates* should have on its upper side a channel 5ft long, an inch wide, and 1½in deep. Water should be poured in, and if it touches the lips of the channel equally, we shall know that the position is level. The reader of Archimedes' works will object that true levelling cannot be made from water, because he argues that water is not level but has a spherical surface, the centre of the sphere being the centre of the earth. But whether water is level or spherical, the extreme ends of the rule should hold the water up evenly. If it is tilting at one end, which is higher, the channel will not have water up to the tops of the lips. Wherever water is poured in, it is bound to protrude in the middle and have some curvature, but the left and right ends should be level with each other. A drawing of the *chorobates* will be given at the end of the volume.

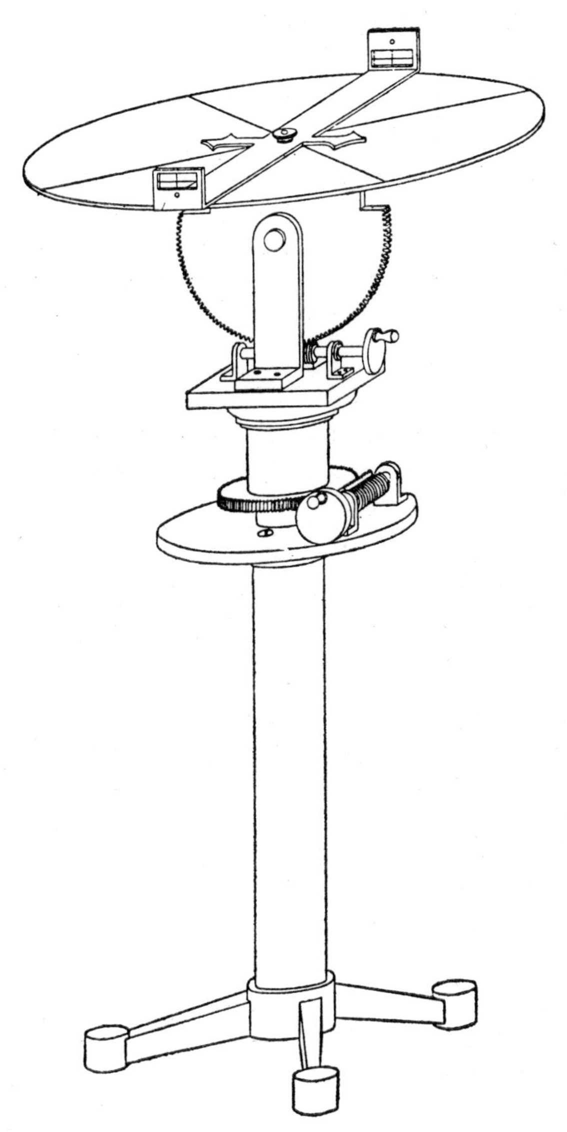

19. Reconstruction of the *dioptra* of Hero of Alexandria, which could be used both for surveying and for astronomical observation

This drawing is missing in the MSS of Vitruvius, but it has easily been able to be reconstructed, since his explanation is very lucid. There is a parallel to the *chorobates* in an ancient Chinese instrument.[11] For more general levelling, such as concerned the *agrimensores*, it was obviously too unwieldy. Instead, hand levelling instruments, the carpenter's square (*norma*) and the plumb-line level (*libra*, *libella*) were used with a levelling-staff. On the *libra* (*libella*), which resembled the miner's triangle of the nineteenth century, the plumb-line covered a vertical line drawn across the front of the cross-bar, and sights were taken on the upper surface of the cross-bar.

'DIOPTRA'

Another instrument not mentioned in the Corpus is the *dioptra* (Fig 19). We have an account of this instrument by Hero of Alexandria, who claimed not to have invented but to have perfected it. Unfortunately there are eight pages of the MS of Hero's *Dioptra* missing, containing the end of the description of the *dioptra* proper and the beginning of the description of the water-level. But we can reconstruct both with certain portions conjecturally supplied. Contrary to the *groma*, it will here be described from the bottom upwards, since that is Hero's order.

Either of two instruments could be fitted to the top of a base (Fig 20) resembling a miniature column. What the bottom of this looked like, neither the text nor the MS miniature tells us. Near the top of the base was a circular bronze disc, on which rotated a hub with a toothed wheel. The teeth of this were engaged by a screw which was set between small brackets fixed to the disc. Accurate adjustment to the rotation of the upper part, whether it was *dioptra* or water-level, could be made by turning this screw. The base was provided with plummets, to ensure that it was upright.

The first of the two upper parts was the *dioptra* proper. At the bottom of it was a cylinder which fitted by means of pegs into the upper side of the toothed wheel on the base. Above the cylinder was a plinth 'like a Doric capital, for the sake of appearance', as Hero puts it. His words show that the MS miniature, besides being sketchy, is inaccurate, since that makes the capital look like a Corinthian one. On the top of the plinth were two tall brackets, carrying between them a toothed semi-

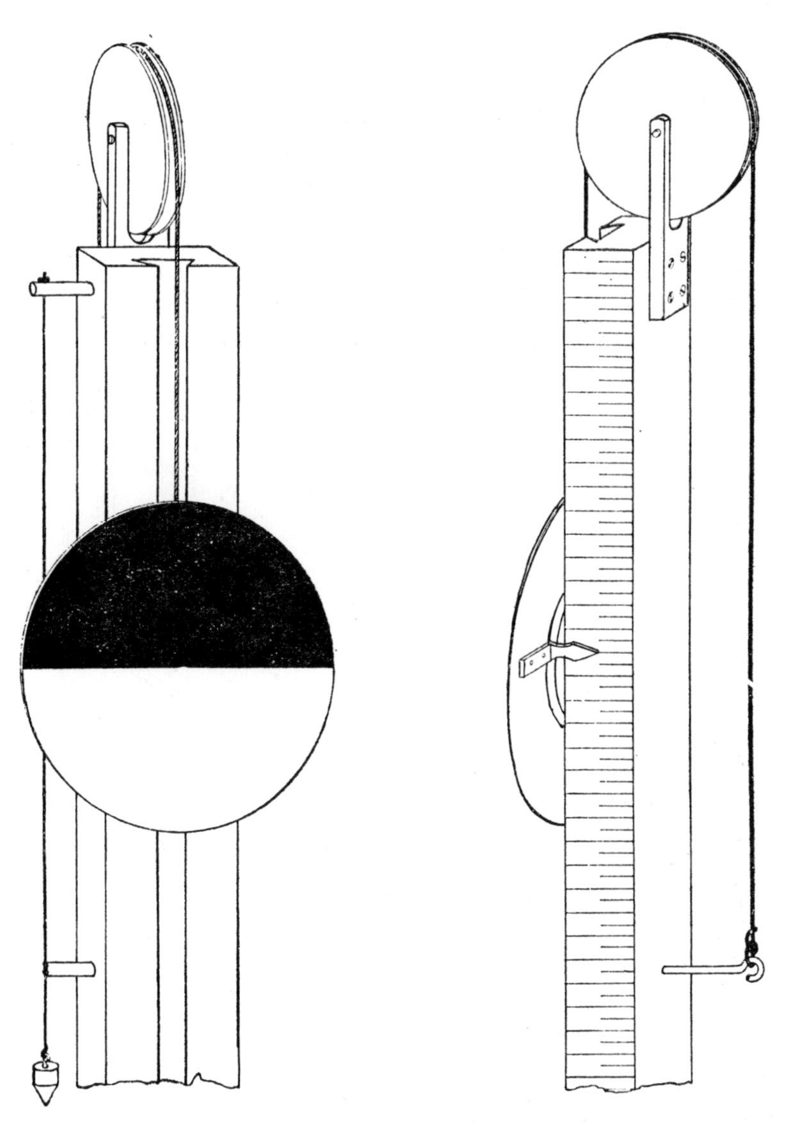

20. Reconstruction of sighting device on measuring rod designed for use with Hero's *dioptra* (front and side views)

circle. Engaging with the teeth of this was a screw like the one on the base, revolving on small brackets fixed to the top of the plinth. This enabled the semicircle and the table on top of it to be revolved, so that the table would tilt upwards or downwards at the required angle. This table was circular, and its top surface was divided into quarters and subdivisions of these. Revolving freely on the table was the sighting-rod, with sights at the two ends.

This instrument, as Hero says, could be used either for surveying, whether of land, buildings, aqueducts, tunnels or harbours, or for astronomical observation. With it one could carry out far more complicated surveying than with the *groma*. Yet although Vitruvius recom-

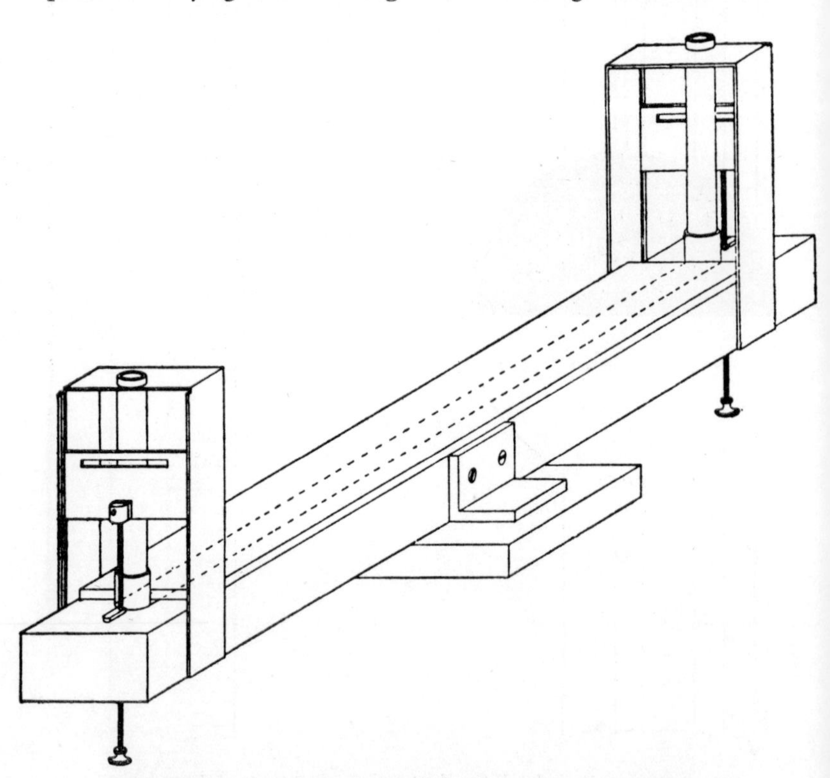

21. Reconstruction of Hero's leveller. This operated by means of a water-level, and could be attached to the column in place of the working part of the *dioptra*

mends a *dioptra*, with some reservation, as an alternative for the levelling of water-courses, and although the elder Pliny recognised its efficiency for astronomical observations, it is never mentioned in the Corpus and was presumably regarded as too elaborate, expensive, and unwieldy for regular use.

Alternatively (Fig 21), on Hero's base one could fit a more accurate water-level than the *chorobates* without having to have an instrument 6m long. Set on a cylinder carrying a plummet (neither shown), a wooden rod about 2m long contained a pipe, to the ends of which were fitted glass tubes. This part was probably Hero's own invention, since it must have depended on glass-blowing, which was invented only in the first century BC. The glass tubes were encased in wooden uprights, slotted into which were bronze plates with intersecting sighting-slits. Each of these two plates could be moved up or down by a screw. For levelling, two measuring-poles carrying plummets were held by assistants, one on each side of the water-level. Each had a half-white, half-black disc (Fig 20), which could be raised or lowered by a cord. The assistants would call out the readings on their poles to the surveyor, who would record the difference in height between these.

HODOMETER

Hero also describes a hodometer (Fig 22), consisting of a box on wheels. By means of four screws turning toothed wheels, each at right angles to the next, a rotating pointer on the top of the box is made to turn very slowly round a marked circle. When the wheel connected to the pointer has progressed 15 teeth, the user will have walked 800 cubits, ie 2 stades. This at least sounds practical, whereas the one in Vitruvius,[12] with round stones placed in drum openings so that one stone fell every mile, is too good to be true. Apart from the Heath-Robinson sound of the contraption, Vitruvius' arithmetic will not work out unless the text is emended.[13]

CONCLUSION

It will be seen, then, that the Roman land surveyor did not go in for elaborate scientific equipment. Although Hero's inventions were probably made in Nero's principate, they did not visibly affect the practical

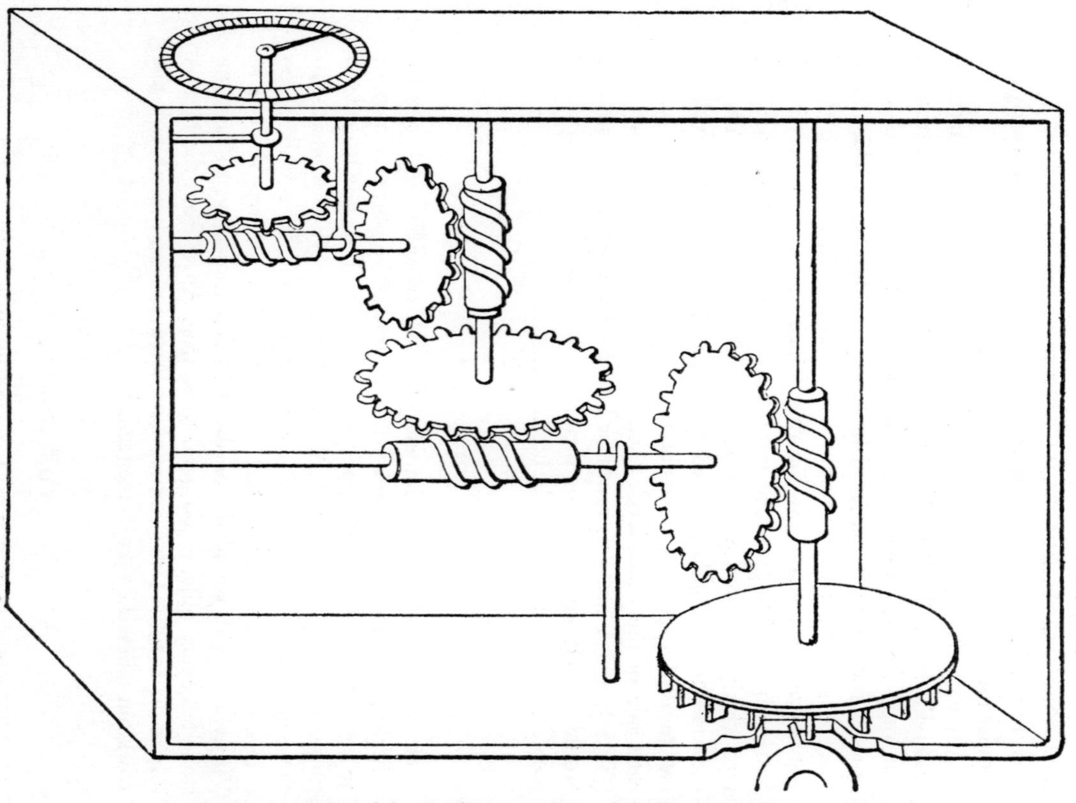

22. Reconstruction of the mechanism of Hero's hodometer (description p. 70)

work of centuriation and boundary measurement which were the main concern of *agrimensores*. If these had their *groma*, measuring equipment, plumb-line level, sundial and writing and drawing aids, they thought themselves for practical purposes adequately supplied.

6

Measurement and allocation of land

ROMAN SURVEYING MEASUREMENTS

Both in town and country, in civilian and in military use, the basic measurement was the Roman foot (*pes*). Many ancient nations worked from parts of the body, so that it is not surprising if we find parallels between Egyptian, Greek, and Roman measurements.[1] The normal Roman foot was 11·6in/29·57cm, but in addition there was an early foot of over 29·73cm and a *pes Drusianus* of 33·3 or 33·5cm, and from the third century AD a shorter foot of 29·42cm. For measurements of buildings and distances in towns, normally only the foot was used. Military surveyors were geared to the usages of the legions, which reckoned to a large extent in paces (*passus*). The *passus* was in fact a double pace, right foot to right or left to left, and this was reckoned as 5 Roman feet. For longer distances the mile, which derives its name from *mille passus*, 1,000 paces, was used, this giving 5,000ft to the mile. Of the various companies that constructed the Antonine Wall in Scotland, some used feet as the unit and some used paces.

The chief measurement of length used by Roman land surveyors for plots was the *actus* of 120 Roman feet. At the standard foot this gives a length of 35·48m. There is an approximate Greek equivalent, the *schoinos*, 120ft of 27·77cm, recorded on the Heraclea tablets. *Actus*, literally 'a driving', was in origin like many other Roman institutions an agricultural term, indicating the distance that oxen pulling a plough would be driven before turning. The sides of 'centuries' were always reckoned in multiples of *actus* (Fig 23). So if traces of squares are discovered with sides which are not multiples of *actus*, one can be reasonably sure that they are not part of any Roman centuriation, except perhaps for 220½ *iugera* (p 85). The Oscans and Umbrians did not use an *actus* but a *versus* of 100ft, and although the word *versus* literally means 'a turning' we should understand it as having the same meaning as *actus*—we may compare *versus* in the sense of a line of prose or verse—but with a shorter span.

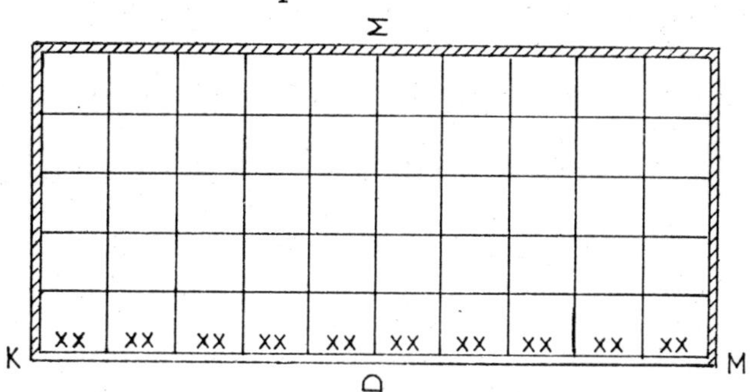

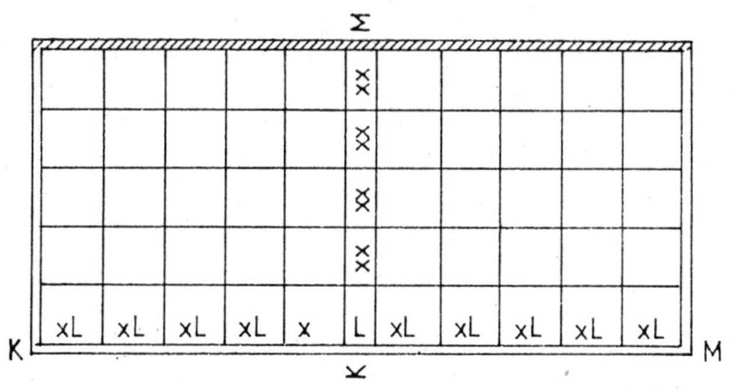

23. Centuriated areas which have rectangles instead of the usual squares: (a) *decumani* 40 *actus* long, *kardines* 20 *actus*; (b) vice versa (K in the centre is the scribe's mistake for D). Hyginus Gromaticus

The chief measurements of area used by the land surveyors were:

1 sq *actus* (*actus quadratus* or *acnua*) = 14,400 sq ft (Roman) = ca 0·312 acres/0·126ha

2 sq *actus* = 1 *iugerum* = 28,800 sq ft (Roman) = ca 0·623 acres/0·252ha

2 *iugera* = 1 *heredium* = ca 1·246 acres/0·504ha, heritable plot

100 *heredia* = 1 *centuria* = ca 124·6 acres/50·4ha, 'century', of regular size

In addition to these, we find a very large number of subdivisions of the *iugerum* and of the *uncia*, which was one-twelfth of a *iugerum*. Columella uses these for his measurements, but most of them are hardly encountered in other authors. When Hyginus Gromaticus wants to denote $\frac{2}{3}$ by abbreviation, he uses the abbreviation for $\frac{1}{2}$ followed by the abbreviation for $\frac{1}{6}$, instead of Columella's term *bes* used for $\frac{2}{3}$ *iugerum* as well as $\frac{2}{3}$ *as*.

The two most commonly used measurements of area in Roman land survey were the *iugerum*, which we can think of roughly as $\frac{5}{8}$ acre or $\frac{1}{4}$ha, and the 'century', roughly 125 acres or 50ha. The *heredium*, which as Varro says was originally enough for one man, must have been important in early times as a small-holding, 100 of which grouped together gave the 'century' its name; but in course of time it disappeared from use. The *iugerum* (connected with *iugum*, 'yoke') was in origin an agricultural measurement, the amount that could be ploughed in one day. Pliny[2] says: 'It is a fair day's work to plough 1 *iugerum* 9in/23cm deep for the first time, 1$\frac{1}{2}$ *iugera* the second time. If the soil is hard, $\frac{1}{2}$ *iugerum* can be ploughed the first time, 1 *iugerum* the second time.' Columella[3] is wrong in saying: 'Formerly the "century" was so called because it contained 100 *iugera*'; he has forgotten about the *heredium*.

The normal size of a 'century' was a square 20 × 20 *actus* (2,400 Roman feet square), giving a size of 200 *iugera*. One may suspect that after the Augustan Age this size became so standard that exceptions to it are virtually non-existent. Over huge areas of the Po valley and Tunisia no centuriation except in the regular 200-*iugera* 'centuries' is to be found, and they are extremely common elsewhere too. They are given by the *Libri coloniarum* as the size for the great majority of areas covered. It is therefore possible that in some areas, where from ground or air survey a non-standard size has been suggested, the squares are non-Roman or some mistake has been made by the investigators.

Non-standard 'centuries' are indicated from three sources:

(a) from the Orange inscriptions, Survey A. The inscribed stones of this survey have rectangular, not square lines round each 'century'. The largest total recorded in a 'century' of this survey is 330 *iugera*, and we may take it that the size was 400.

(b) from the Corpus:

50 *iugera*. Some in Italy founded by the Second Triumvirate (43–33 BC).

200 *iugera* irregular, 25 × 16 *actus*. Beneventum (mod. Benevento), Velia and Vibo Valentia. Siculus Flaccus[4] tells his readers to watch carefully for these, giving Beneventum as an example.

210 *iugera*, 21 × 20 *actus*. Cremona, as stated by Frontinus and Hyginus Gromaticus and confirmed by observation.

240 *iugera*, 24 × 20 *actus*. Aeculanum (le Grotte, near Mirabella-Eclano).

400 *iugera*, 40 × 20 *actus*. Augustan allocations at Veturia near Emerita (Mérida, Spain).

640 *iugera* (?), 80 × 16 *actus*, an unlikely size given by the *Liber coloniarum* for Luceria (Lucera).

The 'century' of 50 *iugera* gave sides of 10 *actus* each to form a square; all the others were rectangular.

(c) from observation in the field:

72 *iugera*, 12 × 12 *actus*, perhaps at Ad Tricesimum (Tricesimo), belonging to Iulium Carnicum,[5] and at Forum Iulii (Cividale, north of Aquileia).

180 *iugera*, 20 × 18 *actus*, perhaps at Belunum (Belluno).

210 *iugera*, 21 × 20 *actus*. Cremona, as above, and Aquinum (Aquino), which also, however, seems to have a system of larger rectangles.

220½ *iugera*, 21 × 21 *actus*. Acelum (Asolo) and Tarvisium (Treviso). These, if correct, constitute the only fractional number of *iugera* in a 'century'.

250 *iugera*, 25 × 20 *actus*. Perhaps on the island of Pharos (Ital. Lesina, mod. Hvar).

300 *iugera*, 30 × 20 *actus*. Altinum (Altino), thought by Fraccaro to be 40 × 30.

Measurements of 'centuries' have not revealed as uniform a standard as might be expected. With the regular Roman foot of 29·57cm, which gives the *actus* as 35·484m, the normal length of 20 *actus* should measure 709·68m. The shorter foot recognised from the third century AD gives 20 *actus* as about 706m. In fact the average in Italy works out at about 707m (704–6 and 710–14 in Campania, 705 in Romagna, just over 710 at Eporedia (Ivrea), 711 in Emilia). In Tunisia the average is about 708m; but in the area of the Shott Jerid, Barthel calculated it as 703m. The low measurements in Africa no doubt reflect the third-century reduction in the size of foot. The variations in Italy, where centuriation is likely to be earlier, may simply be due to inexact local measures or lack of care in measuring.

ORIENTATION IN PRACTICE

Many ancient races aimed at exact orientation by compass points in planning towns, monuments or the countryside. Many ancient temples, though not necessarily Roman ones of the classical period, were planned exactly to face the four compass-points. The Roman augur, before observing the auspices, had to ensure that he was facing south or east.[6] The surveyor likewise had to take his bearings before conducting a new survey, and his orientation and terminology were to a great extent, as will be seen, based on those of augury. Military survey was governed by similar rules. When no considerations of terrain arise, Roman camps tend to be orientated towards the four points of the compass. The beginning of the treatise on the building of camps, which probably contained instructions on orientation, is unfortunately missing.

Surveyors had to be familiar with methods of fixing due south (p 56), but they do not seem always to have taken the trouble to calculate it exactly. If a full-size sundial was available locally, this could be brought into service. Alternatively, for a rough calculation, they could use a portable sundial (p 70). It was not uncommon for adjacent centuriated lands to have a slightly different orientation, which might be due to a variety of factors, one of which may have been the use of as fallible an instrument as a portable sundial. Hyginus Gromaticus has been quoted as castigating those who take bearings from the rising or setting sun, and insisting that centuriation in a colony shall be started from a correct south bearing.

The main lines of survey were, in general, either exactly or approximately north–south and east–west. For establishing right and left, near and far sides (see below), the surveyor might face any of the four compass-points. Frontinus, the earliest extant writer on surveying as such, prefers west, which certainly ties up with the idea, favoured by him and the grammarian Verrius Flaccus, that the *kardines* should run north–south, the *decumani* east–west. But the maps in the Corpus usually have east at the top, a practice very common in antiquity and the Middle Ages, and Hyginus Gromaticus regards this as the most correct orientation. Neither account corresponds fully with fact, since both ignore the centuriation in which the surveyor faced north or south. In practice the commonest direction for him to face was east, south being the second preference. East also possesses antiquity, since Gracchan centuriation stones confirm it. It is most significant, and does not seem to have been appreciated by scholars, that these two directions are the same as those which augurs faced. As Thulin saw, there was obviously some continuity from Etruscan religious observances. (It has been suspected that Varro, in his *Roman Antiquities*, invented certain alleged Etruscan antecedents; but there is no evidence that any of these were connected with surveying.) The Corpus (*Liber coloniarum*, 209L) quotes the Campanian land as having its *kardo* towards the east and its *decumanus* towards the south. This, as Castagnoli remarks, is probably due to the existence of wider north–south roads in that area. In the three Orange surveys (p 163ff) the surveyor faced respectively east, west and west, if we agree with Piganiol's reconstruction.

In the minority of cases where there is a great divergence from the four cardinal points, it is normally due either to existing main roads or to geographical features. Thus at Tarracina (Terracina) the *decumanus* was the Via Appia; in much of the Po valley it was the Via Aemilia; at Acelum (Asolo) it was the Via Postumia. In the area round Florence the topography, particularly as regards rivers, dictated the alignment of the centuriation, whereas the streets of the city faced the four cardinal points. In some coastal areas the alignment of the coast affected the centuriation; thus much of the Dalmatian coast runs roughly NW–SE, and at Iader (Zara) the centuriation was made to conform to this coastline, even spanning coastal indentations, as shown by Bradford.[7]

DIVISION OF LAND

(1) *Centuriation.* The usual Latin name for dividing land into 'centuries' was *limitatio*. But since 'limitation' has another sense, we derive our word from the alternative name *centuriatio*; though 'delimitation', likewise derived from *limes*, would perhaps be equally apt. *Limes* properly means a boundary zone; in agriculture, a path or balk. Where the technical writers speak of a straight boundary line having no width, they use instead the word *rigor*. Each of the main *limites* in a centuriated area was either a *decumanus* (*decimanus*) or a *kardo* (*cardo*). The main streets were the *decumanus maximus* and *kardo maximus*, abbreviated to DM and KM. The literal meaning of *cardo* is 'hinge', but it was used also for a pole of the earth or vertex in the sky, so that it is appropriate

enough for a north–south *limes*. As to *decimanus*, it literally means 'belonging to the tenth' (*decimus*). The precise application of the word to its use in centuriated areas is disputed (see Appendix A), but to the present writer it seems most likely to mean 'very big', even though this interpretation results in tautology in the phrase *decumanus maximus*. Perhaps in origin this sense comes from the Roman notion that every tenth wave is the biggest. Since the *limites* between 'centuries', where they existed, normally served as roads or farm-tracks of various widths, we can for practical purposes speak of them as such. But it is important to remember that the Romans, with their agricultural background, thought of them primarily as boundaries having a certain width. Surveyors spoke of *limites prorsi*, those straight ahead of them or behind them, and *limites transversi*, those at right angles to these.

Centuriation was normally applied to *ager publicus* (p 178), land acquired by the State through conquest and largely cultivated by individual lessees or sub-lessees. This is likely to have taken place on a limited scale from the late fourth century BC onwards. When Tiberius Gracchus had regulated the maximum holdings of *ager publicus*, it was ruled that the Gracchan land commission should have centuriated any which had not earlier been, after which this came to be the regular practice. Since colonies were normally established on *ager publicus*, centuriation came to be commonly associated with colonies. But we find exceptions to these associations: thus in some areas the territory of *municipia* or even smaller places came to be centuriated.[8]

Whenever land was to be divided out, the *groma* was set up at a point where the main intersection (*tetrans*) was to be established. This could be in different types of location, as shown below.

(a) *At or near the central point of a settlement*. This layout is discussed on p 122 ; it was neither common nor, as Hyginus Gromaticus implies, borrowed directly from military camps. It seems to have been relatively popular in North Africa, since in addition to Ammaedara, the example quoted by him, it has been noted at Thubunae (Tobna, Algeria) and Hr Zouda, Tunisia.[9] From Italy it has been recorded at Allifae (Alife) and Luca (Lucca).

(b) *On an existing road in a settlement*. This is particularly true of Tarracina (p 115), where the colony was set up on the line of the Via Appia, which then became the *decumanus maximus* of the settlement.

(c) *At a point not far outside the existing or planned settlement*. This was done where the surveyor wished to make a fresh start on unimpeded terrain.

(d) *At a distant point.* This applied particularly in cases where mountains, rivers, etc, impeded nearby centuriation.

A formal initiation of the survey required the participation of the colony's founder or his representative and an augur. The founder would be a magistrate under the Republic, the emperor or a member of the imperial family under the Empire, while the augur would take the auspices, and if they were favourable the founder or his representative would duly authorise the start of the survey. In observances of this sort, as well as in orientation, we may see the lingering influence of Etruria.

The first line to be sighted with the *groma* and laid out was the *decumanus maximus*. It would be the widest street: in the case of Augustan colonies for veterans, the width was 40ft. Then the *kardo maximus*, 20ft wide in Augustan colonies, was sighted and laid out at right angles. Distances were measured with 10ft rods, and cross-checks were taken to ensure accuracy of angles and measurements. The correct procedure was for centuriation stones to be laid at all crossroads. But even allowing for the high casualty rate, the small number remaining suggests that in many places only a selection were laid. It seems quite possible that some crossroads were given wooden markers. The stones at the main intersection were inscribed with the abbreviations DM and KM (Fig 24).

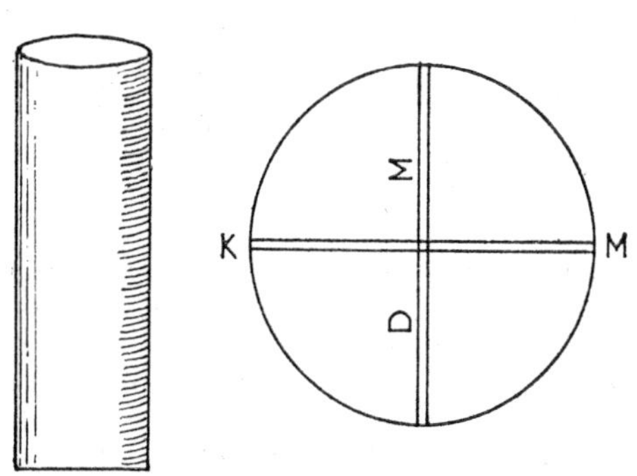

24. Centuriation stone, marked on top with lines at right angles and KM (*kardo maximus*), DM (*decumanus maximus*). Such a stone would have stood at the centre of the centuriated area. Hyginus Gromaticus

If the 'centuries' were to be square and of normal size, a distance of 2,400 Roman feet would be marked off in each direction from the main intersection. Other *limites* were then sighted and laid out at right angles, care being taken to make a cross-check every time. These *limites*, if they were the main ones next to the central pair, could be labelled D II or K II; the same numbering evidently served, as far as they were concerned, for each direction.* But as opposed to the 'century' numbers, the street numbers were very often not inscribed. As a result some writers in the Corpus, presumably composing their manuals under the later Empire, are unsure about the numbering of *limites*, and wonder if there was a D I between DM and D II. Surely there never was: the idea probably arose out of the numbering of 'centuries'.

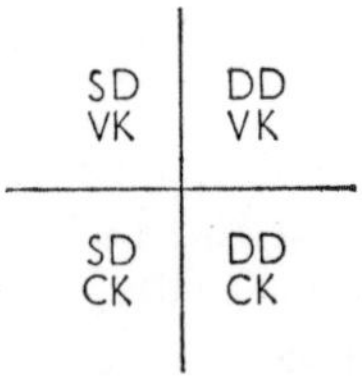

SD VK	DD VK
SD CK	DD CK

25. Latin abbreviations for the four main areas of a centuriation system: left and right, in either case in front of or behind the surveyor as he started his survey

The method of numbering these is shown in Fig 25. Each 'century' was given a pair of co-ordinates, according to the following table:

DD	*dextra decumani*†	To right of DM
SD	*sinistra decumani*	To left of DM
VK	*ultra kardinem*‡	Beyond KM
CK	*citra kardinem*	Near side of KM

* This point seems to emerge from the Corpus and its diagrams, but Warmington (1967) takes a different standpoint. In his diagram on page 174, (a) he says that at the intersection one block north of the centre one would expect four stones inscribed SDIKM, whereas the present writer's view is that the inscription, which in practice might have been only on one stone, should be DII KM; (b) three blocks west of the centre he would give DMKKIII, while the present writer would give DM KIV or DM KIIII; (c) he shades a 'century', in the north-east corner of which a stone presumed to read DMVKVII would have been placed. This refers to *Corp. Inscr. Lat.* I^2 639, on his page 168, and on this the only enumeration is KVII. Here the full inscription would surely have been DM KVII.

For CK we sometimes find the less correct KK. When the surveyor stood at the crossroads and faced the original orientation (Fig 26), the 'century' immediately in front of him on his right was DD I VK I, and the rest were numbered as in Fig 27. This system of nomenclature extended unbroken as far as the centuriation went, so that quite high figures are sometimes attained. Owing to the regularity of the chess-board system, it is sometimes possible, from the discovery of only one or two centuriation stones *in situ*, to try to work out where the centre of the centuriation must have been. This has been done in Campania (p 144), in areas of the Orange cadasters (Chapter 11), and in Tunisia (p 156), but in no case beyond doubt.

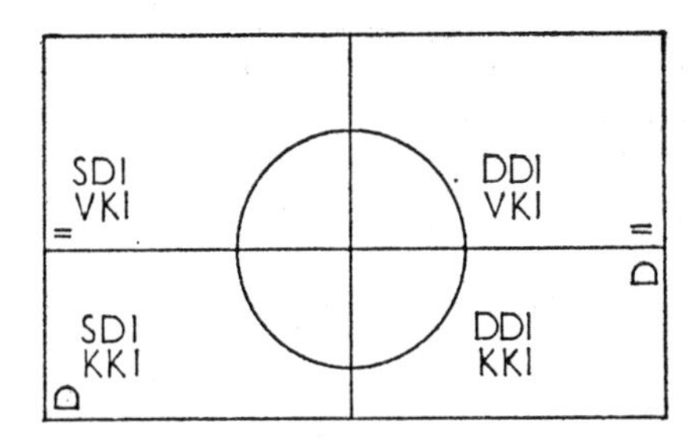

26. The four central 'centuries', a miniature illustrating Hyginus Gromaticus's treatise on *limites*

In the original survey the number of each 'century' was supposed to be recorded on a centuriation stone. Hyginus Gromaticus gives what appears to be the correct way of doing this. Each century would theoretically have at three of its corners, the corner nearest to the centre of the survey and the two adjacent to it, street numbers. Thus on these three corners of 'century' DD I VK I (Fig 27) would appear

Where the enumeration of 'centuries', rather than *limites*, is involved, there is no disagreement among scholars.

† The usual modern expansion of DD and SD gives with Hyginus an incorrect accusative form *decumanum*. Whereas *ultra* and *citra* are prepositions taking the accusative, *dextra* and *sinistra* are adjectives in the ablative feminine, 'on the right/left (hand)'.

‡ The terms *antica* and *postica* (p 32) seem to have been used in early times for the regions *ultra* and *citra* (not vice versa, as given by Rudorff in Blume *et al*, ii, 341).

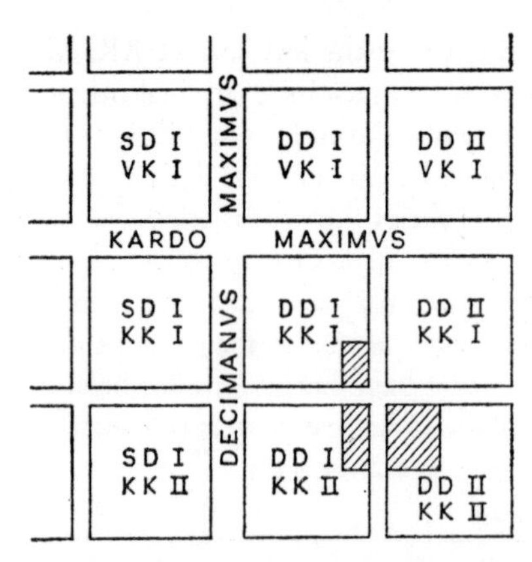

27. The method of numbering 'centuries'. The shaded area, which might belong to a single owner (an exceptional example quoted by Hyginus Gromaticus), lies within 'centuries' DD I KK I, DD I KK II and DD II KK II ($6\frac{2}{3}$, 15, and 45 *iugera* respectively)

DM KM at the centre of the survey, DM on the *decumanus maximus* at the end of the first 'century', and KM on the *kardo maximus* in a similar position. This, if adhered to, left only the fourth side of the 'century', that farthest from the centre, free to receive a stone recording the 'century' number. Hyginus Gromaticus recommends that it should be written vertically, presumably because the amount of space for a horizontal inscription was limited and Roman numerals could well be lengthy. The example he gives is DD LXXXXVIII VK LXXV, for the 'century' 98 to the right of the DM and 75 beyond the KM.

Centuriation stones have been found in many different places.[10] On stones of the Gracchan land commissioners, the title of the commission is given, eg *tresviri agris iudicandis adsignandis* or in one case *tresviri agris dandis adsignandis iudicandis*, in the abbreviated forms A.I.A. or A.D.A.I., followed by the names of the three commissioners and the 'century' number. These Gracchan stones are cylindrical, with a slight taper and a diameter of about $1\frac{1}{2}$ft at the top; they stood about 2ft up from ground level. Gracchan stones have been found as follows: one at S. Angelo in Formis (p 144), one in the territory of Suessula, one near Sicignano degli Alburni (east of Eboli), two near Sala Consilina in the territory of Atina, one at Polla nearby, three near Rocca S. Felice in the territory of Aeclanum. A Gracchan stone restored at a later date has been found near Fano, and there is another possible example near Carthage. The other important group of centuriation stones is that belonging to the scheme of AD 29–30 carried out in Africa Nova by the third legion Augusta (p 156). Many of these have very large numbers, in one direction up to 280 'centuries', about 198km, from the centre.

The *limites* did not all have the same width. The figures given above for the two main streets are fairly typical, except that in some areas the KM too was 40ft wide. Entries for particular colonies in the *Libri coloniarum* contain the phrase *iter populo debetur*, 'the highway is due to the people', followed by a number of feet varying from 10 to 120. This was rightly explained by Saumagne[11] as referring to a legal servitude of so many feet on the main road in a colony. The only oddity is why, if the texts are sound, there was such an exceptional range in the width of the servitude. Every fifth main *limes* was made wider than the intervening ones, to ensure that it became a usable road; such a *quintarius* might have a width of 20ft (Fig 28). The ordinary main *limites* were often 12ft wide, and there was a prescribed minimum of 8ft. It

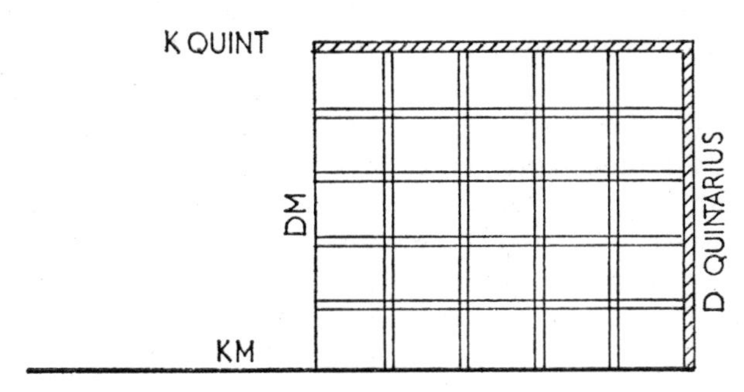

28. Centuriation, showing in each direction a *quintarius*, fifth *limes* which was wider than the others: Hyginus Gromaticus

was stipulated that they should be kept free of undergrowth and open to the public. Where they were in a built-up or frequented area, no difficulty about this law normally arose, but the example quoted from Frontinus on p 124 shows that this was not necessarily so elsewhere.

There could be internal *limites* within a 'century'. These are called by the Corpus *limites intercisivi*, literally 'balks which cut inside'. In several instances we find a single division in each direction, dividing a 'century' into four equal parts, each of which in the case of the standard size contains 50 *iugera*. Another method of dividing into four is by rectangular strips, each 20 × 5 *actus*, as at Padua. Similarly at Asolo, where the 'centuries' are 21 *actus* square, they are divided into three strips of 21 × 7 *actus*. In North Africa the commonest subdivision[12] is into 5 × 10 = 50 rectangles, each measuring 480 × 240ft and containing 4 *iugera*. In some 'centuries' two or more of these have not been separated, so that a group contains 8, 12 or even 16 *iugera*. We also find 'centuries' divided into 20 rectangles of 10 *iugera*, with 3 × 4 internal *limites*; and others divided into 50, 100 or 144 rectangles, this last being obtained by means of internal *limites* at 200ft intervals.[13]

(2) *Subseciva*. As Frontinus tells us, *subsecivum* literally means land cut away by a line. There were two types.

(a) If, in centuriating land, any areas remained between 'centuries' and the outer boundary of the land, they would officially be classed as *subseciva*. Unofficially, provided they were a fair size, they might be known as 'centuries'. An example from the Orange inscriptions is the land immediately west of the Rhône. The illustration accompanying Frontinus' text is shown on p 101.[14] The miniature below it is intended to show an area near the R. Pisaurus (p 107).

(b) If, within a 'century', there was either land unsuitable for allocation or a lack of landholders, some part could be designated a *subsecivum*. We learn from the Corpus that land could have the same legal status as *subseciva* if it was within a centuriated area but not allocated.

Some *subseciva* were sold (or absorbed, evidently with taxation) by Vespasian, some presented to their occupants by Domitian. An example of the latter is recorded in an inscription[15] of Falerio in Picenum dating to AD 82. Domitian ruled in this case that *subseciva* originally allocated to inhabitants of Firmum, a neighbouring town, should, as the town of Falerio requested, be granted to their then occupants (*possessores*).

(3) *Strip Allotments*. There were three types, all rather uncommon: (a) allocation by *decumani* alone; (b) *strigae* and *scamna*, which may be thought of as brick-shaped parcels of land; (c) narrow strips (*laciniae*). Allocation by *limites* running in one direction only is similar to the system used in the Greek colonies of Southern Italy. Something of this sort has been observed at Cosa in Etruria,[16] 16 *actus* wide, where more pattern has emerged than is to be seen on the general plan in Salmon,[17] at Alba Fucens, 12 *actus* wide, and at Cales.

The Latin names *strigae* and *scamna* were originally agricultural terms, the former meaning a swath or windrow, the latter a bench or balk. As far as land survey was concerned, they designated rectangles of land, the former bounded lengthwise as viewed with the orientation chosen, the latter breadthwise. This strip system was an old one, but continued under the Empire, sometimes alongside centuriation. Frontinus remarks that arable lands in the provinces were in his day cultivated in such rectangles.

There is much discrepancy over the shape of such holdings. In Fig 29a, illustrating Frontinus, they look like long narrow strips round a rectangular field; in a miniature illustrating Hyginus Gromaticus, they

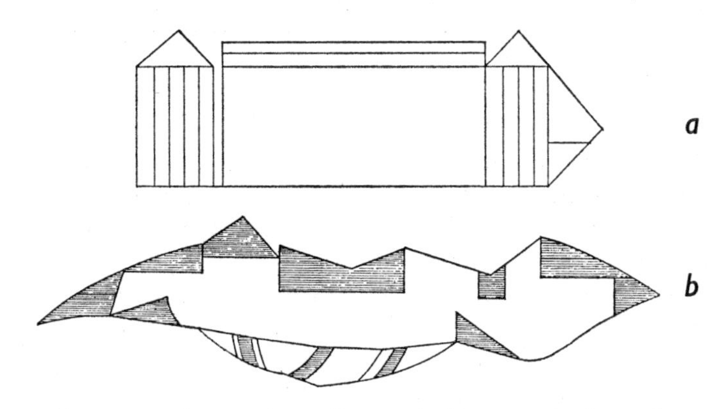

29. (a) *Strigae* and *scamna* (strip allotments); (b) private land, merely measured by surveyors and not centuriated: Frontinus

are horizontal rectangles of various sizes, whereas the text makes it sound as though horizontal and vertical rectangles were interspersed. There is also uncertainty over *limites*. Frontinus says: 'Parcelled and allocated land is that of colonies. This has two types: one mostly contained within *limites*, the other allocated by straight lines bounding adjacent properties, as at Suessa Aurunca in Campania. All land which according to this type is bounded lengthwise is called *strigae*, breadthwise *scamna*.' On the other hand Hyginus Gromaticus speaks of roads 12ft wide enclosing in one direction two *scamna* and one *striga*. As to areas, Nipsus gives land in *scamna* as 24 × 20 *actus*, 240 *iugera*, but Weber thinks the wording of Hyginus Gromaticus[18] suggests rather 30 × 20 *actus*, 300 *iugera*.

Similar 'bricks' of land occurred in Cyrenaica, where in 96 BC the Romans inherited from Ptolemy Apion royal lands of that shape measured in Ptolemaic feet.[19]

(4) *Unsurveyed land.* Apart from measured land, centuriated and otherwise, there was land known as *arcifinius*. This was in origin frontier territory, and was subject not to civil but to international law. Frontinus, *De agrorum qualitate*, tells us that it 'is bounded, following long-standing observance, by rivers, ditches, mountains, roads, trees in front, watersheds, or ground occupied by another landholder earlier' (p 101). In *De controversiis* he speaks of 'all lands, with curved boundaries, called *arcifinii*, such as at ends of a ploughed field or promontories or mountain tops or river courses or any projecting areas'.

ALLOCATION OF LAND

The allocation of centuriated land was one of the important duties of a land surveyor. When the initial stages of the survey had been completed, he would divide the required number of 'centuries' into holdings of the right size. In cases where the holdings in a 'century' did not add up to the exact number of *iugera* in it, a settler could be given holdings in adjacent 'centuries', such as those shaded in Fig 27. The surveyor decided the order of tenure of holdings by lottery, for which he used an urn. Lottery was a common practice in antiquity, used not as today for gambling but in many spheres of politics, law and even (at the temple of Fortune in Praeneste) fortune-telling. As far as land allocations were concerned, it was recognised at least as early as the Gracchan period, from 133 BC onwards. The main idea was to avoid disputes, since certain areas would be or would come to be more valuable or useful than others; grumblers like the legionaries whom Tacitus quotes could easily find poor holdings to complain of. It might well come about, on the other hand, that part of a centuriated area was built on, and a block (*insula*) might become valuable. In such cases settlers might attempt after a while to sell their holdings. The laws prohibiting alienation of holdings on *ager publicus*, apart from remnants, tended in course of time to be ignored. Whether surveyors acted as valuers when lands were sold, we do not know.

The exact procedure of lottery is differently given by Hyginus, *De Limitibus*, and by Hyginus Gromaticus.[20] The former has a vague account, not specifying more than one lottery and evidently starting from the centre of the survey. The latter, on the contrary, is specific and starts with as many 'centuries' as are required from the boundary of the land available. He assumes, as an example, that each settler is to receive 66⅔ *iugera*, one-third of a 'century'. In that case the names of settlers will be grouped in threes and their order will be determined by lottery; they will constitute 'Entry 1', 'Entry 2' etc in his ledger. Then, by a second lottery, the 'centuries' will be pulled out of the urn, and if DD XXXV VK XLVII comes out first, the surveyor will write in his ledger:

Entry 1. DD XXXV VK XLVII
Lucius Terentius son of Lucius, tribe Pollia, 66⅔ *iugera.*
Gaius Numisius son of Gaius, tribe Stellatina, 66⅔ *iugera.*
Publius Tarquinius son of Gnaeus,[21] tribe Terentina, 66⅔ *iugera.*

Finally we should note that the surveyor actually had to take new settlers in person to their allotted lands, so that there should be no mistake over the location or extent of these. While all this was going on, he had an obligation to make entries in his ledger. Only when the whole of the allocation was completed would he proceed to the registration and mapping, as described in Chapter 8.

7

Boundaries

ROMAN BOUNDARIES AND THE LAND SURVEYOR

Mommsen[1] was perhaps not over-simplifying when he wrote that basically there were only two concerns of the Roman land surveyor: land allocation and boundaries. There is no doubt that boundaries played a most important part in their routine, and this importance is reflected in the amount of space devoted to the subject in the Corpus.

Boundary marks were common throughout antiquity. We have seen that they were in regular use in Mesopotamia from the very earliest historical times. In some areas of the Mediterranean and Near East natural boundaries, such as rivers, were generally recognised; in others there were man-made boundaries. Where the latter served to separate properties, they were more often walls than hedges. Where neither of these types of boundary was available, boundaries were marked either by land-marks (including trees) or stones or both.

With the Romans, boundary stones (*termini*) had certain religious connections.[2] There was a god Terminus, at whose festival, the Terminalia, on 13 February, members of a neighbourhood sacrificed to him and feasted in the open near one or more boundary stones. Whereas centuriation stones had no religious significance, it was the correct practice for each boundary stone or group of boundary stones to be set up with due ceremony. An animal was sacrificed, and its ashes and other offerings were placed in a hole dug for them. Many of these customs are enumerated by Siculus Flaccus.[3] Dolabella is quoted in the Corpus[4] as saying that the woodland god Silvanus was the first to plant a boundary stone, and that every estate has three Silvani, one to guard the house, one for the country as shepherd's god, and one called *orientalis* ('eastern'), who has a grove on the boundary. To move a boundary stone without permission was considered a religious as well as a civic offence.[5] The religious connotation of boundaries is also seen in the instructions to the surveyor by Hyginus Gromaticus that in some areas he will have to set up stone altars, recording on the side facing a colony the boundary of that colony, on the other side the name of the neighbouring townspeople. Where the boundaries of three authorities met, the altar was to be triangular (p 102, upper plate).

The land surveyor could find himself involved in boundary questions both in the official and the private sectors of his practice. On the official side, in addition to surveying, parcelling and allocating land, the *mensor* had to see that its boundaries were properly established and marked. The technical terms for this were *terminatio* or *determinatio*. Areas between the edge of the centuriation and the boundary of the settlement would be surveyed as *subseciva* (p 94). Then the outer limit of the settlement would be fixed from natural boundaries or by boundary stones and recorded. On private estates the surveyor might well be invited in only where a boundary dispute arose.

From the legal point of view, it was important to lay down exactly how far, in the topographical sense, jurisdiction extended.[6] On the larger scale, every province had its boundaries with Italy, if adjacent, and with other provinces. The outer boundary of the Roman Empire was known by the same term, *limes*, as we find so commonly used in surveys. But there is a difference: in the case of a colony, for example, the *limites* were all within its territory, and its boundaries were called *fines*. Italy itself was divided by Augustus into eleven regions.[7] In the following list some of the less important tribes are omitted:

1 Latium, Campania.
2 Apulia, Calabria.
3 Lucania, Bruttii.
4 Samnites, Marsi, Paeligni, Sabines etc.
5 Picenum.
6 Umbria, Ager Gallicus.
7 Etruria.
8 Gallia Cispadana.
9 Liguria.
10 Venetia, Istria.
11 Gallia Transpadana.

These divisions persisted with little alteration down to the time of Constantine. They were used by geographers, encyclopaedists, including the elder Pliny, and the *Libri coloniarum*.

The extent of jurisdiction of a new colony was normally provided for by a definition of the *territorium* in the charter. Thus territory and jurisdiction went hand in hand, and the surveyor needed to know enough of the laws regarding local spheres of jurisdiction, as well as be able to define the territory, to make what decisions were required. Surveyors would be brought in, among other things, whenever there was an extension of *territorium*.

The boundaries either of a local authority (Fig 30) or of individual estates were in classical times written up in the customary formula depending on successive points. Hyginus[8] gives the formula thus: 'From

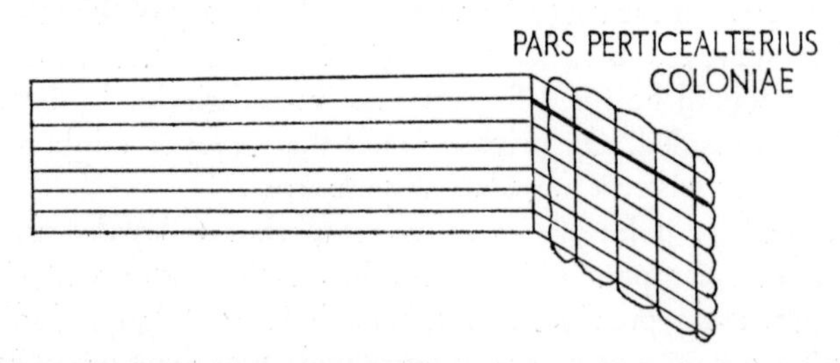

30. Boundary area between two differently aligned systems of centuriation (*pars perticae alterius coloniae*, 'part of the other colony's survey')

the hill called A to river B and over river B to stream C or road C, along that road to the foot of mountain D at the area called E, along the ridge of that mountain to the summit and over the summit by the watershed to the area called F, then down to area G and from there to the crossroads of H and so back to the starting-point.'

An early practical example is contained in an arbitration decision of 117 BC[9] defining the boundary between the hilly land of the *castellani Langenses* near Genoa and the lower land of the Viturii: 'There is a boundary stone at the confluence of the Edus (Edis) and the Procobera. From the R. Edus (Edis) to the foot of Mt Lemurinus, where there is a boundary stone. Then straight up the ridge of Mt Lemurinus; there is a boundary stone on Mt Procavus', and so on. We may compare the details of the boundary disputes between Delphi and Anticyra and between Lamia and Hypata (references p 43).

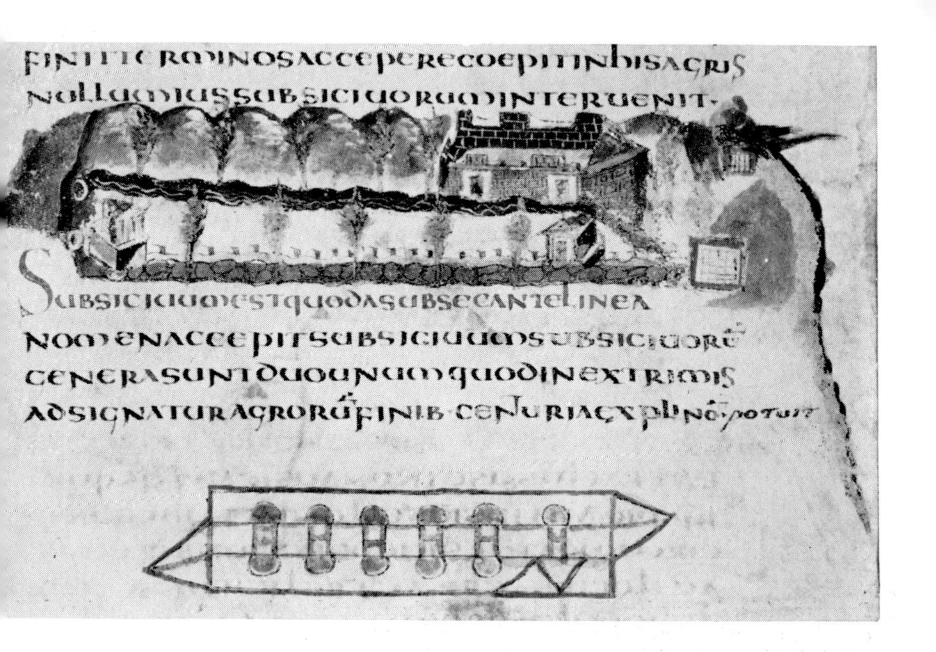

e 101 (*above*) Upper miniature: unsurveyed land, including farmhouse. Lower miniature: *seciva* (remnants of land). Translation of text (Frontinus): 'A *subsecivum* derives its name from a : that cuts away (*subseco*). There are two types of *subseciva*: (1) areas which will not make up a ntury'' in allocated lands.' The text goes on: '(2) inside allocated lands and within whole cen- es'. From MS A, Wolfenbüttel; (*below*) areas of *subseciva* near the River Pisaurus, Italy. These nants of land here lay within a 'century', but were downgraded as subject to flooding. Miniature (Agennius Urbicus) from MS A, Wolfenbüttel

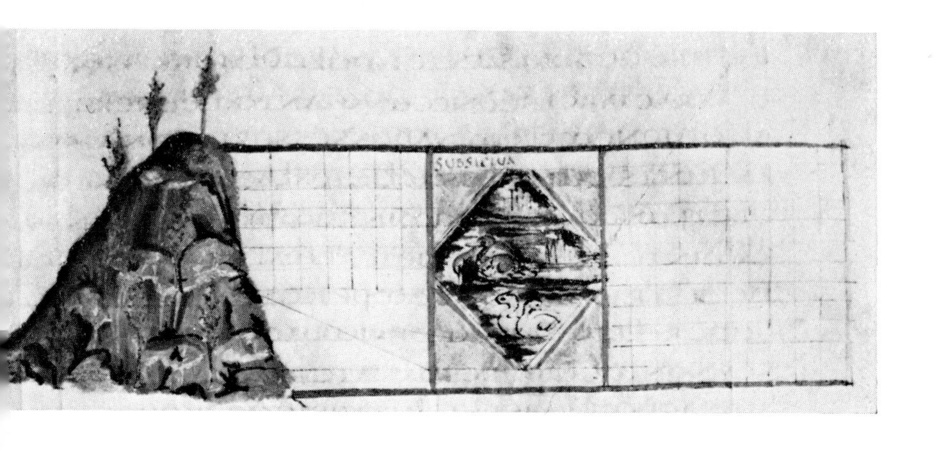

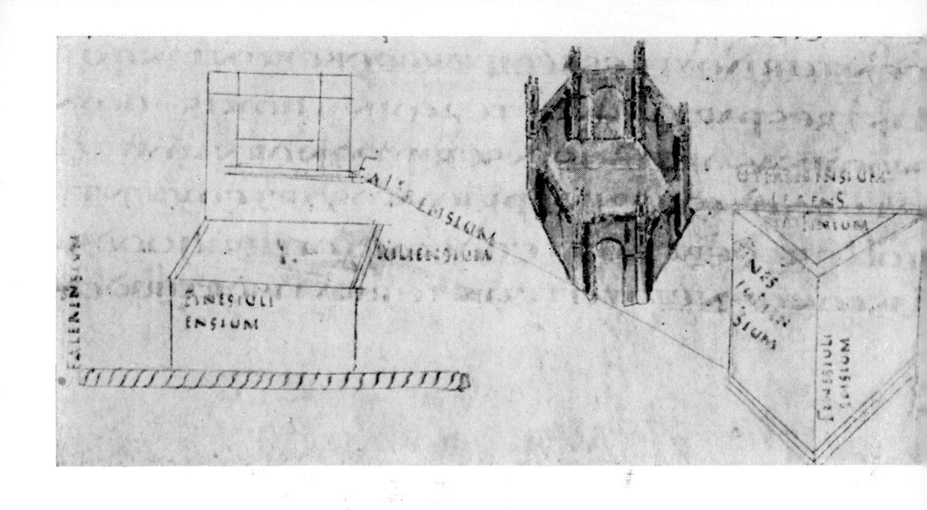

Page 102 (*above*) Boundary lines and a triangular boundary mark between the territories of adja colonies, exemplified as Iulienses (inhabitants of any Colonia Iulia) and Falerenses (inhabitan Falerio in Picenum). Miniature (Hyginus Gromaticus) from MS A, Wolfenbüttel; (*below*) bound between adjacent farms (F = *fundus*, 'farm', followed by adjective from or genitive of owı name). The preceding text concerns boundary disputes between neighbours. Miniature (Fronti from MS A, Wolfenbüttel

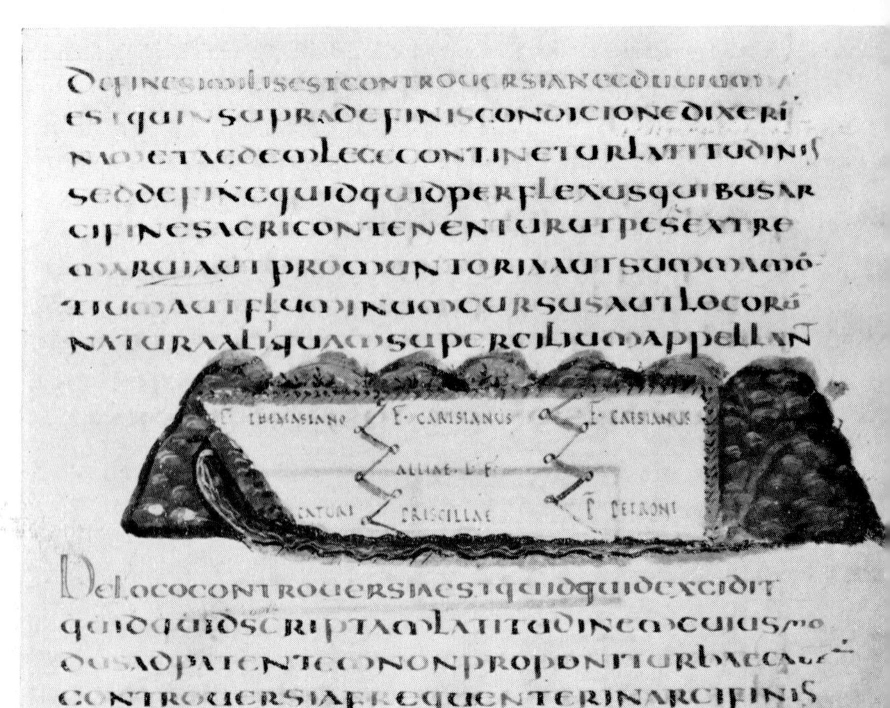

Examples of cylindrical boundary stones from the Gracchan period and later are given in Warmington.[10] They include the abbreviation FPV (*fundus possessoris veteris*, 'farm of earlier occupant') and others of less certain meaning. They include restrictions, eg: 'The lower road is private property belonging to Titus Umbrenius son of Gaius. Entry on foot by permission; no cattle or carts.'

The inviolability of boundary stones was guaranteed under the Lex Mamilia of the late Republic. Augustus ruled that boundary stones should be of flint or other stone, and that they should be set 2½ft in the earth and protrude 1½ft. Under later emperors boundary stones tended to be square in section, not cylindrical.

Compass points are encountered under the late Empire. Specimens are to be found in the Albertini tablets, forty-five wooden tablets written in ink and discovered in 1928 near the frontier of Algeria and Tunisia.[11] This is an area which with the Vandal incursions gradually lost the advantages of Roman civilisation and was eventually overwhelmed by desert. The tablets refer to sales, mainly of property, between AD 493 and 496, and are in African vulgar Latin, except for legal phrases, for which mainly correct Latin is used though sometimes incorrectly spelt. An extract may be rendered: 'Likewise elsewhere an area called Pullatis, in which there are five olive trees between neighbours of the same area, on the east Quintianus, on the south Quintianus, on the south-west and north-west Victorinus Nugualis.' The signature is that of an agent on behalf of the illiterate vendors.

A late treatise in the Corpus, the *De terminibus* (= *terminis*), illustrates and describes a large variety of boundary stones, and also gives a good idea of the types of natural boundary adopted. These included groups of trees, as in the preceding example, and even isolated trees, which might be thought to be too fallible. A reproduction of two of the treatise's pages was given by the author,[12] with attempted translation in the accompanying article. A typical entry is: 'If a boundary stone is hewn square and has a dot on its side, it indicates a spring (Fig 31). But if it has a hollow on top, it indicates a well at the boundary.' If we are to believe the writer of the treatise, different boundary stones had to be set up not only where there was a three or four-way boundary, but where there were certain prominent features of a property, such as a water-tank, a spring, a stream or running water, a well, certain trees, mountain features etc. It is clear that the writer has chiefly in mind large private estates. The preoccupation with water is very natural in the Mediterranean, where vast areas may, unless irrigated, lack water for many months of the summer. A spring might dry up during the summer months, and the special boundary stone would then indicate its whereabouts.

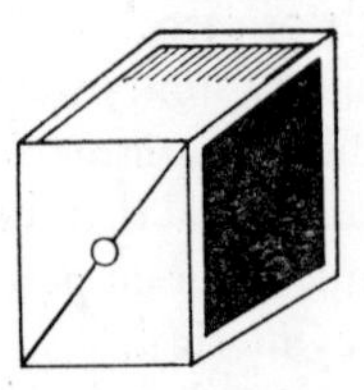

31. Boundary stone indicating the presence of a spring: Latinus

Cryptographers may care to try their hand on several sets of abbreviations which appear in Blume.[13] The first two are said to have been inscribed on successive boundary stones in the territory of Volaterrae (Volterra) and 'in various territories in Italy, especially near the R. Nemus'. The latter starts: AI. AM. IN. KM. IK. DI. KO. MX. XM. YP. FI. HO. SV. VS. ZE. QP. PT. HN. GY. AB. CO. GH. RV. LM. RM. QP. VS. TV. On p 357 of Blume this text appears in slightly different form, and the explanations are mostly (apart perhaps from KM) unhappy guesswork:

prima	linea	prima	mensura	prima	norma quae nota
A	L	A	M	I	N

primum	decumanum	k	m	primum	k	decumanum	primum
I	D	K	M	I	K	D	I

The writer of this is betrayed by his spelling, since he explains H as *hortogonium*, which should be spelt *orthogonium* and thus not start with H at all. On pp 363–4 of the same volume will be found attempts to explain single letters found on boundary stones.

Access to fields was carefully protected by law. One of the laws passed at Julius Caesar's instigation was the Lex Mamilia, of which the Corpus preserves Chapters 53–5. Mommsen[14] claimed it as part of Caesar's agrarian legislation of 55 BC; but Rudolph argues that it was not an agrarian law *in toto*; if so, it perhaps belongs to Caesar's dictatorship. Among its provisions was a guaranteed minimum width of 5ft for all access roads leading to neighbours' property. This had the effect of imposing a legal servitude, either on one landholder or on two neighbouring ones. All such access ways had to be kept clear of undergrowth; likewise ditches and waterways had to be kept clear. Penalties for offences against these regulations and for removing or changing the position of boundary stones are laid down. Frontinus was writing about 140 years after the Lex Mamilia, but clearly these provisions were very much in force.

BOUNDARY DISPUTES[15]

Like boundary stones, boundary disputes go back to remote antiquity. In featureless landscapes it was rather easy to move stones by night and try to annex some of one's neighbour's land. 'Thou shalt not covet thy neighbour's house' applied equally to his estate. Much litigation, of which the ancients tended to be fond, was expended on such questions.

We have seen (p 51) that Roman land surveyors were often called in to solve, or arbitrate in, boundary disputes. The contestants might be private individuals, local authorities, such as those in Greece already mentioned, or the State. As the conditions of tenure varied from place to place, the surveyor had to be familiar with the law in so far as it affected boundaries and boundary disputes. He would act in two capacities: (a) as legal adviser on any points arising, (b) as arbitrator or judge in a restricted sphere, that of determining where the boundary lay and what was the legal position of any land within 5ft of it. Procedure parallels from German and Swiss law are quoted by Rudorff.[16] Whenever this second procedure was adopted, the surveyor-judge and both disputing parties, with their supporters, had to go to the area in dispute. Ovid[17] recommends his reader not to make a present of disputed land but to say clearly 'This is my land, that is yours'; here and elsewhere we find the poet's early legal training brought out in his verses.

We possess Frontinus' treatise *De controversiis*, together with a commentary on it by Agennius Urbicus. He starts by saying that disputes may be over boundaries or over areas, but adds that further subdivision is needed; though in fact we find that these two reappear among his fifteen types of dispute. These are not entirely logical, as they arise from cumulative headings under which legal action can be taken, and there is inevitably some overlapping.

(1) *Position of boundary stones.* In the absence of hedges or fences it was

in many places easy for unscrupulous people to move boundary stones, as indicated above. This was expressly forbidden by the ancient Twelve Tables, codified in the Lex Mamilia of Julius Caesar's dictatorship. Frontinus stresses the importance of seeing that all owners of immediately adjacent land are consulted.

(2) *Straight line boundaries.* The word *rigor* means, in surveying parlance, a boundary consisting of a straight line or (if the plural) straight lines. Wherever a dispute was over land within 5ft of such a boundary, it was dealt with under the Lex Mamilia; again the original law, specifying 5ft as the distance, was in the Twelve Tables.

(3) *Other boundaries.* Here again the rule including all disputes within 5ft applied. This category included curved boundaries, where there might, for example, be a dispute over the exact curve.

(4) *Position.* This arose whenever anyone encroached on adjacent land, being particularly common in the type of land called *arcifinius* (p 96), where the boundary often consisted of a variety of landmarks. In more settled areas, the dispute would fall within the following section.

(5) *Area.* This often occurs in centuriated and assigned land. If a landholder could prove by old titles that he was entitled to more lands, then he could be given extra land, for example, in an adjacent century. The principle is that the *limes*, even if it serves as a public road, need not also constitute the outer boundary of one person's holding.

(6) *Ownership.* This type of dispute often occurred where, as in Campania (p 207), a wood belonging to a farmer was a long way from his farm. It sometimes arose in connection with common pastures. Disputes over inheritance might also fall under this heading.

(7) *Occupation.* The Latin term *possessio* did not imply legal possession, but rather what we might call acknowledged occupation. Questions of this sort were largely linked with *ager publicus* (p 178), and limits were set by various laws to the amount of land that could be held by *possessio*; though we do hear of such tenure also on private land ('Lex Thoria', 111 BC). The sources of the Roman law of *possessio* are far from clear. But the services of a surveyor might obviously be needed in any civil case where *possessio* was contested.

(8) *Alluvial land.* When river action disturbed holdings, landholders might or might not have ground for legal redress. At this point there is a gap in the text, but the commentary enlarges on the type of dispute.

In the case of lands occupied but not assigned (§7 above) there is no legal redress. But in the case of centuriated and assigned lands, there could well be a claim against another landowner or against the community. The areas particularly mentioned by the commentator as affected by river action are Gaul and the R. Padus (Po); and he records that the R. Pisaurus (Foglia) was allotted in the survey the fullest width it ever attained (p 101). Another area quoted is that of Augusta Emerita (Mérida), founded by Augustus for veterans and given an enormous territory. Its Roman remains today are among the most conspicuous in Spain, but no centuriation is to be seen. According to the Corpus most of the veterans were settled not near the R. Ana (Guadiana), but on the outskirts. Even after three allocations of land, some land remained unallocated. The usual method of selling off remnants (*subseciva*) was adopted. But as they felt it to be unfair that anyone should buy part of a river, which would strictly be considered public property, the governor of Lusitania was persuaded to designate a specific width for the river.

(9) *Territorial rights.* Two local authorities might have a dispute over the extent of their jurisdiction, as far as it affected either town or countryside. This jurisdiction would vary according to the status of the local authority, whether colony, municipality or small urban district (*conciliabulum*). A case involving a detached area is quoted from Interamnium Praetutianum in Picenum.

(10) *Remnants* (*subseciva*). Encroachments on land not assigned within a century, or on land outside the centuriated area but within the territory of a colony, would come under this heading.

(11) *Public woods, pastures, etc.* Lands could be marked on the surveyor's map as 'public woods and pastures' of, say, the people of Augusta Concordia (Beneventum). If they were so marked, even without the word 'public', they could under no circumstances come under private control. The term 'public woods' did not necessarily mean that the public was allowed access to all parts: for example, the local authority might grow timber and cut it for fuel to heat public washhouses or for other such purposes. The authority also ran poor people's cemeteries, execution areas, etc.

(12) *Excluded lands.* These, outside the surveyed area, came under the same laws as remnants (*see* 10).

(13) *Land belonging to religious cults.* Disputes over this were normally

dealt with under ordinary law, only needing a surveyor if a question of measurement arose. It is interesting to observe from the commentary that, as Christianity grew under the Empire, some Christians occupied land formerly (or perhaps still officially) belonging to pagan cults and sowed it.

(14) *Rain-water*. Disputes about rain-water were common in antiquity. A case in Athens in the fourth century BC hinged round the question whether a certain stretch was torrent or path. Parallels in the Mediterranean today, especially with barrancas in Spain, can easily be found. Cases of complaint about the diversion of rain-water frequently occurred; but there was no need for a surveyor to be brought in unless the question of boundaries arose.

(15) *Roads*. Theoretically all *limites* in a centuriated area were public rights of way. In practice, on difficult terrain, farmers tended to plough them up or plant trees on them.

(16) *Fruit on trees*. Overhanging produce could be picked by the adjacent landowner. There might be disputes whether it was in fact projecting.

Boundary disputes were by no means always so trivial. The emperor himself not infrequently intervened to appoint a surveyor-judge in important boundary disputes between local authorities. The institution continued well beyond the fall of Rome, as shown by the episode recounted by Cassiodorus (p 45).

THE CANNAE STONE

A stone re-used in a modern building 2km from Cannae (Castagnoli, 1948) and dating from AD 76 records that the emperor Vespasian restored the boundaries of the public lands of M.C. (presumably the *municipium* of Cannae) in accordance with public maps; the *f* and *m* of *formis*, 'maps', are probable restorations. This is to be linked with Vespasian's action on *subseciva*, as recorded on p 94. The type of map, assuming the existence of these, will have been similar either to surveyors' maps (Chapter 8) or to the Orange cadasters (Chapter 11).

8

Maps and mapping

The Romans used maps for various purposes. There were world maps; maps illustrating geographical treatises or works of literature; road maps and itineraries to help travellers find their way about; official and military maps; detailed town plans; and surveyors' maps. Other technical maps and plans are likely to have existed but have not survived.

The most famous map of the Roman world, serving in some measure as propaganda, was the one set up as the result of Agrippa's collection of material. Pliny[1] says that although Agrippa compiled the preparatory material, the map was started by his sister after his death (12 BC). On completion by Augustus it was set up in a colonnade in the Campus Martius. Copies of it were probably erected elsewhere: Eumenius (third/fourth century AD) pleads for what was probably such a copy to be placed in a portico at Augustodunum (Autun). For the remoter countries Agrippa's map probably relied on the world map of Eratosthenes (ca 275–194 BC). This has not survived, and the maps illustrating Ptolemy's *Geography* (second century AD)[2] survive only in medieval copies. The extent to which these are likely to go back to Ptolemy has been disputed, but it is generally thought that Vatopedi 754 (twelfth or thirteenth century), several other Greek MSS and one Latin (Vaticanus lat 5698) are descended from Ptolemy's maps. One of the early Virgil MSS, Vaticanus lat 3225 of ca AD 420, has maps of the Aegean islands, not named, and of Sicily, to illustrate the text of the *Aeneid*.

Although we do not possess an original Roman road map, we have in the Peutinger Table[3] a medieval version of one. It is a long strip of parchment preserved in the National Library at Vienna, and may be the *mappa mundi* which the monk of Colmar says he drew on twelve skins of parchment in 1265. It originally depicted the Roman world in twelve segments, of which the most westerly, containing much of Britain as well as parts of Spain and North Africa, has unfortunately perished. The curious feature of the Table is its elongation: it is 6m 74 long and only 34cm high. The most likely interpretation of this is that in its original form it may have been on a papyrus roll, in which case the distortion may have been an original and intentional element. Other explanations are: (1) that the parchment of the original necessitated this elongation; that is unlikely, as parchment books were mostly

32. Rough map of Britain indicating centres of administration zones under the late Empire: Maxima Caesariensis; Valentia, Britannia prima; B. secunda, Flavia Caesariensis. Notitia Dignitatum

made up in codex form, not in rolls; (2) that it was derived from the shape of Agrippa's portico map, assuming that this was not a globe (*orbis terrarum* = 'earth'); (3) that it is connected with Ptolemy's idea that in map-making one should stress known areas at the expense of unknown.

The Notitia Dignitatum[4] illustrates its list of military and civil officials under the Late Empire with very inadequate maps, of which we have medieval copies. Fig 32 gives a rudimentary map of Britain towards the end of the fourth century AD, with five walled cities as

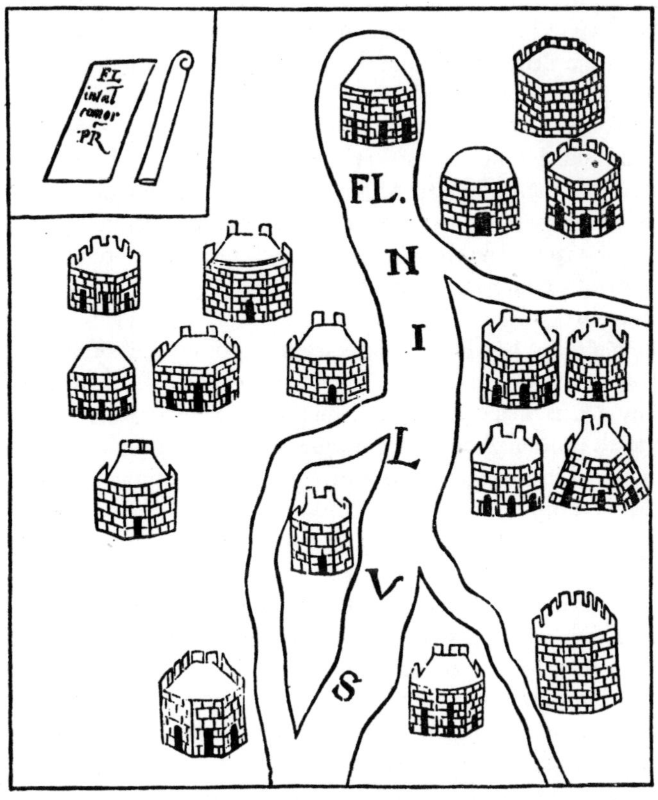

33. Map of Egypt. Key (in the MS spelling): Filari, Hermunthus; Hermipolis, Lato; Tentyra, Cussa; Oasis Minor, Asfinis, Thebas, Apollonos, Syene maior; Presentia, Diospolis, Lyco, Copto; Copto (sic), Ambos, Maximianopolis. Notitia Dignitatum

capitals of the five subdivisions of Britain at the time. Fig 33 gives somewhat more detail of the Thebaid in Egypt. We do not possess a Roman military map, but Roman soldiers in the East may well have used maps compiled in Greek, and we have one such, the Dura Europus shield map. This parchment shield cover, unearthed at Europus on the Euphrates in 1923, contains the name of the Danube in two forms, Istros and Danubis, and names various stations along the west coast of the Black Sea. In addition to maps, the military clearly used itineraries simply listing places and distances. Examples of these are the so-called Antonine Itinerary (probably third century AD) and the Ravenna Cosmography (about AD 670). An attempt to plot on the map the names in the British section of the Ravenna Cosmography is made by Richmond and Crawford.[5]

The Forma Urbis Romae[6] is a marble plan of Rome compiled soon after AD 200. Its numerous surviving fragments may be seen in the courtyard of the Capitoline Museum in Rome. Although its measurements are not wholly accurate, it may be said to operate basically on a scale of 1:300. The drawing up of such plans was presumably the function of architects, not of land surveyors; and the plans of the Agrimensores can seldom have been on such a large scale.[7]

Mapping was, however, an important part of the theory and practice of Roman surveying. The Latin for a surveyor's map is *forma*, which literally means 'shape' and is also, among other things, the word for an architect's plan of a house. In the sense of a surveyor's map it is already encountered in the agrarian law of 111 BC and continues in use to the end of the Empire. Surveyors' maps may be divided, somewhat arbitrarily perhaps, into two classes: (1) working maps, (2) teaching maps. Our knowledge of the first class has to be pieced together from various sources, including adaptations for teaching purposes. For the second class we have what purport to be copies of the originals; some of these illustrate texts which are teaching apprentices about working maps. The teaching maps sometimes contain the word *ut*, 'as', showing that only an example is being given. Clearly we should be wrong in thinking that every teaching map is based on a working map, though there must have been some overlap between the categories.

(1) *Working maps*. It was part of the duty of a land surveyor to make a map (*forma*) of any land he had divided up. These maps were normally on bronze,[8] and unfortunately none survives. The reason for this is

undoubtedly that, with the barbarian invasions and the break-up of the Roman Empire, bronze was regarded as a valuable metal to melt down and take away. Likewise no surveying records on papyrus, other than the Greek ones from Egypt mentioned on pp 27 ff above, and none on parchment or wax tablets have survived. We can, however, form a reasonably good idea of these maps from the Corpus and its illustrations, from the Orange tablets and from the Forma Urbis Romae. It might be thought that metal, though durable, would prove too inflexible for such a purpose, but an interesting comment, showing that this was not necessarily so, comes from Hyginus Gromaticus. If, he says, a settler has, in addition to his holding of centuriated land, a plot of woodland or mountain pasture, this may be recorded by hammering an extra piece of bronze on the edge of the map.

The maps were drawn up in two copies. One of these remained in the local community, while the other was lodged in the Emperor's record office in Rome, the *tabularium*, built in 78 BC. The imposing façade of this office, incorporated into later buildings, is to be seen on the west side of the Roman forum; no documents from it have come to light. Hyginus Gromaticus[9] comments:

> When we have ended all 'centuries' with inscribed stones, we shall surround parts assigned to the State, even if they are centuriated, with a private boundary, and shall enter them on the map appropriately, as 'public woods' or 'common pasture' or both. We shall fill the whole extent with the inscription, so that on the map of the area a more scattered arrangement of lettering may show greater width. We shall likewise bound excepted or granted farms, giving them inscriptions as with public places. We shall similarly show granted farms, eg 'Farm of Seius granted to Lucius Manilius son of Seius'. In Augustus' allocations of land, excepted farms have a different status from granted farms. . . .

Later he comments:

> We shall write both on the maps and on the bronze tablets (*tabulae aeris*) all mapping indications, 'given and assigned', 'granted', 'excepted', 'restored, exchanged for own property', 'restored to previous owner', and any other abbreviations in common use, to remain on the map. We shall take to the Emperor's record office the mapping registers (*libri aeris*) and the plan of the whole surveyed area drawn in lines according to its particular boundary system, adding the names of the immediate neighbours. If any property, either in the immediate neighbourhood or elsewhere, has been given to the

> colony, we shall enter it in the register of assets. Anything else of surveying interest will have to be held not only by the colony but by the Emperor's record office, signed by the founder. This is how we shall allocate undeveloped land in the provinces. But if a borough has its status changed to that of colony, we shall examine local conditions. . . .

As an example of detail he adds later:

> We shall inscribe the lots in such a way that if a single holding extends over two or three or more 'centuries', these 'centuries' shall be inscribed in one single lot with the amount of land held in each. For example, if 66⅔ *iugera* are given to one man and are split between three 'centuries' thus, DD I KK I 6⅔ *iugera*, DD I KK II 15 *iugera*, and DD II KK II 45 *iugera*, a single lot will have to embrace these three. The rest will be carried out according to this example. We shall take the owners of lots to their land and assign boundaries to them. When we have assigned the boundaries and carried out the remaining functions of a surveyor, we shall duly produce the maps and everything else connected with the survey. . . .

Whereas the *tabulae aeris* were perhaps, as their name implies, records on bronze accompanying the maps, *libri aeris* can hardly have been 'books of bronze', the literal translation. Rather they must have been books about the bronze maps, ie ledgers relating to them. Such a ledger must at first have been on a papyrus roll, later either on a papyrus roll or on a parchment or vellum codex; none is extant. The type of detail recorded may have been very similar to those on the Orange tablets, whose form, however, perhaps combines features of the maps and the surveyors' ledgers. Use was made of standard abbreviations for the wording on the maps.

It is clear from the extract quoted above that the surveyor, even in later centuries, had to be familiar with a whole range of land law, especially that dating from Augustus onwards, so as to give each holding its precise legal status on the map. Seius is used, like John Doe, as a legal specimen name. In Blume et al fig 185, if a famous Faustina is alluded to, it will be a reference to the empress or daughter of Antoninus Pius. The reference to Scipio as the person originally granting the farm is obviously intended to be to one of the famous generals of that name. Were it not that it is probably invented as an example, one could quote it as illustrating continuity over 300 years or more.

The holding of 66⅔ *iugera* mentioned in the extract, one-third of a 200 *iugera* 'century', was one in common use for veteran colonies under the Empire. The way in which it is regarded as being allocated is illustrated in Fig 27. If all holdings in a colony were equal, all consisting of a number of *iugera* exactly divisible into 200, and all 'centuries' were completely allocated, there would be no need of such a jigsaw pattern. But even if holdings were equal, in certain places some 'centuries' contained an area of *subseciva* which threw the count out.

The Orange tablets (Chapter 11) differ from surveyors' maps in several ways. In the first place they are confined to centuriated land, whereas surveyors' maps were not. Secondly, they were designed for the local authority to have records for taxation purposes, so that the financial side is emphasised; whereas, if we may judge from Hyginus, it was the legal aspect of land-holdings that was stressed in the surveyors' maps. Thirdly, the latter seem to have been more pictorial, as would be likely where mountain lands and woods were included. Finally, the wording does not seem to have needed quite as much abbreviation as we find on the Orange tablets.

(2) *Teaching maps.* The Corpus preserves not only the type of map described above, which bears some resemblance to what we may imagine the surveyors' working maps to have been like, but a number of plans of towns and surrounding lands designed to illustrate points in the texts of the manuals. Since they do not, for the most part, contain the sort of detail that might be expected in official plans, and since their wording is often of general application, they may be labelled teaching maps. These are of several types.

(a) *The single colony with adjacent lands.* This type of diagram was inserted in the text of the Corpus to provide a visual explanation of certain features connected with centuriation. They are sometimes drawn from actual colonies, sometimes given names like Colonia Iulia which are not specific (see p 123 below), sometimes not connected with any particular area. Thus on p 120 is a small map of the colony at Anxur or Tarracina (Terracina), to accompany the text: 'In some colonies they set up the *decumanus maximus* in such a way that it contained the trunk road crossing the colony, as at Anxur in Campania. The *decumanus maximus* can be seen along the Via Appia; the cultivable land has been centuriated; the remainder consists of rugged rocks, bounded as unsurveyed land by natural landmarks.' The site is a very striking one.[10] It is at the point where the Via Appia, on its way south from Rome, meets the sea. To the left of the diagram may be seen the Pomptine

(Pontine) marshes which the Via Appia crossed, and which despite repeated attempts were not finally drained until 1926 onwards. This is now the Agro Pontino, a fertile area because of layers of fine silt brought down in winter by torrents from the Lepini mountains. Up to the first century AD the Via Appia turned inland to climb the spur on which the Volscian settlement of Anxur and the temple of Jupiter were situated. But later, almost certainly in Trajan's principate, the rock now known as Pesco Montano was cut to a height of 128 Roman feet. Every 10ft down it, enclosed in a rectangle, is a numeral indicating the number of feet from the top; the lowest is CXX. We are told that Trajan resurfaced the road, and he may well have done so with some of the stone cut out from this headland, so as to bring this section of the important highway round the base of it.

The colony of Tarracina-Anxur was founded in 329 BC, probably with the title Colonia Anxurnas, which is virtually the form found on the MS miniature. Livy[11] tells us that, after defeating the Volscians, Rome sent 300 settlers to the new colony, each of whom received 2 *iugera* of land. If we reckon four or five to a family, this would represent 1,200–1,500 settlers and families. The holding, a mere 1¼ acres (0·504ha), would not have sufficed for farming. So it is likely that each settler received in addition common pasture land. The area centuriated was that known as La Valle (Fig 34). The MS miniature shows 'centuries' between the Via Appia and the sea. Some of the centuriated land may have been there, but that area was somewhat marshy in antiquity, and what is visible in the shape of rectangles today lies between the Via Appia and the mountains. Each of the two of these represented today by unsurfaced roads and farm-tracks is roughly equal to two 'centuries', and remains of a third NNE and possibly a fourth ESE of the right-hand rectangle on Fig 34 suggest that the total north of the Via Appia was eight 'centuries'. Since the original allocation amounted only to three 'centuries', more land must subsequently have been centuriated. Hyginus Gromaticus, writing under the Empire, comments as quoted above that the land that can be cultivated has been centuriated.

Two other sites clearly identifiable from the Corpus as individual colonies are Minturnae (3km from Minturno) and Hispellum (Spello). Minturnae in Campania (p 153), famous for its marshes in which Marius hid from the pursuing soldiers, was in origin an Ausonian settlement, but was colonised by the Romans in 295 BC and restored by Augustus.

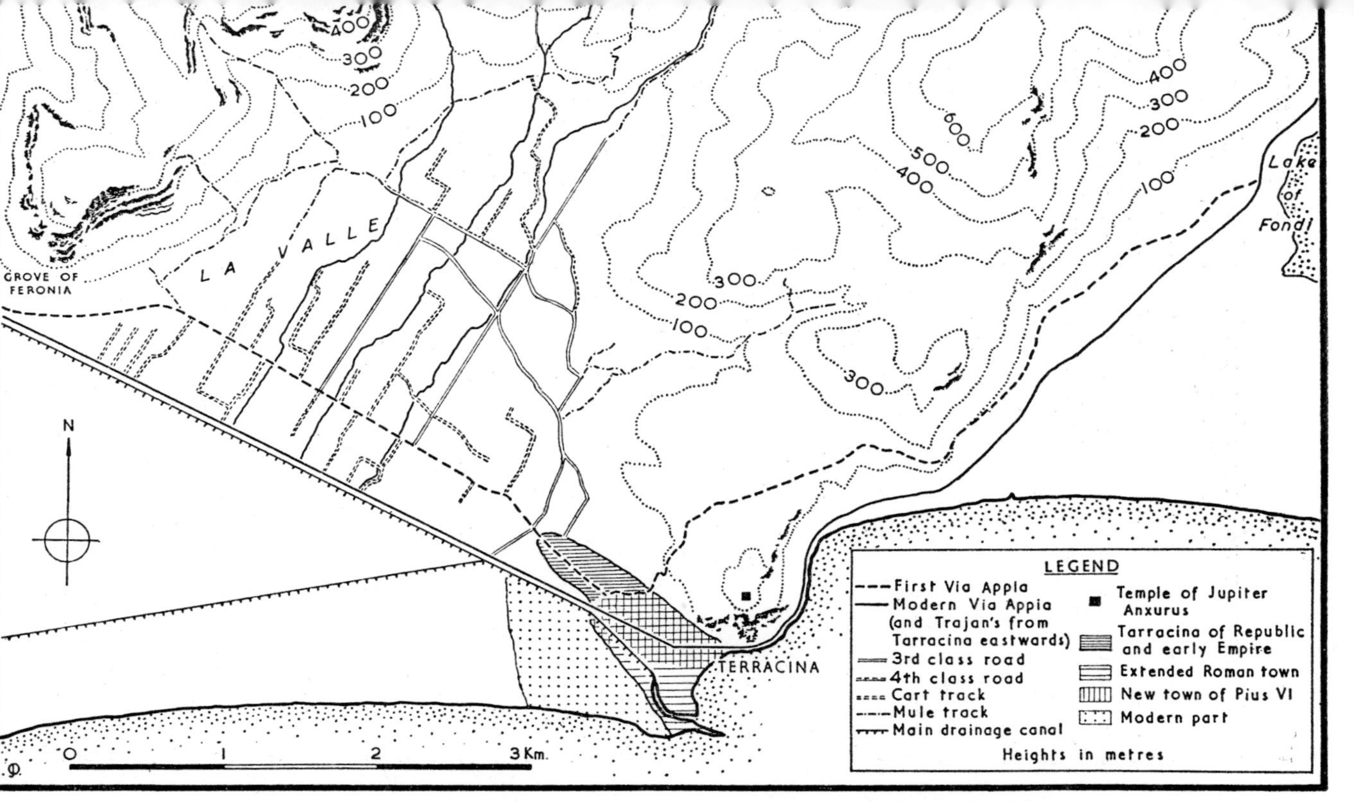

34. Tarracina-Anxur (modern Terracina): La Valle area NW of town, adjacent to Via Appia, showing remains of centuriation

The site was excavated in the early 1930s,[12] but only uncertain traces of centuriation have as yet been found. In the central area of the Roman city there is a small cross hollowed in the stone at what may have been the central road intersection, perhaps used for placing a *groma*. The elder Pliny speaks of Minturnae as being on both sides of the R. Liris (Garigliano), and this is confirmed by Hyginus Gromaticus, who writes in the context of this diagram: 'Augustus also re-founded a number of cities previously founded as colonies but depopulated in the civil wars, by sending out new settlers and sometimes increasing the territory. The result is that in many areas new centuriation cuts into the old at a different angle, the stones at the old points of intersection still being visible. For example, in the territory of Minturnae in Campania, the new assignation on the other side of the Liris is centuriated; on this side of the Liris a later assignation was made on the returns of the previous occupants (*possessores*).[13] Here the original boundary stones have been abandoned.' The MS miniature shows walls on both sides of the R. Liris. The walls of Minturnae, forming a rectangle of 620 × 530 Roman feet,[14] lie only on the right bank of the river, ie the left-hand side of the diagram. The *decumanus* was the Via Appia, with a bridge over the Liris; the *kardo* was formed by the road to Arpinum.

The text referring to Hispellum is also that of Hyginus Gromaticus: 'Many founders (of colonies) looked to suitability of terrain, and set up their *decumanus maximus* and *kardo maximus* where they were going to make the most of their assignation of land. Men of old, because of sudden dangers of war, were not content to wall cities, but also chose rugged, hilly land to provide a natural defence. Such rocky areas could not be centuriated, but were left either as state forests or, if barren, unoccupied. To bring the land of such cities up to the size required for colonies, they were given territory of neighbouring communities, and the *decumanus maximus* and *kardo maximus* were set up on the best soil, as in the territory of Hispellum in Umbria.' Today one can enter the gate of Spello and climb up the steep slopes of the walled town for a good view of the Umbrian plain. The title given on the miniature (p 154), Colonia Iulia, is correct as far as it goes: Hispellum was founded as a colony probably by Octavian. The words *flumen finitimum*, 'neighbouring river', refer to one separating the territory of Hispellum from that of some other town. Modern rivers do not seem to correspond to it, though Schulten[15] thought the R. Ose, bordering the territory of

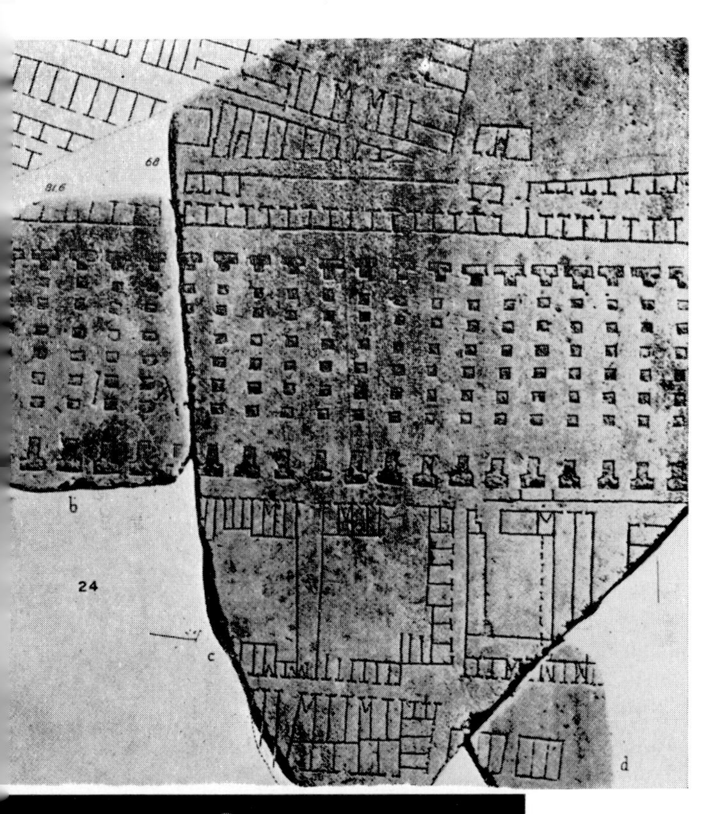

Page 119 (*above*) Fragments of the Forma Urbis Romae, a plan of Rome shortly after AD 200, on a scale of approximately 1:300. In the Courtyard of the Capitoline Museum, Rome; (*below*) two fragments of cadasters of the Roman colony of Arausio (Orange)

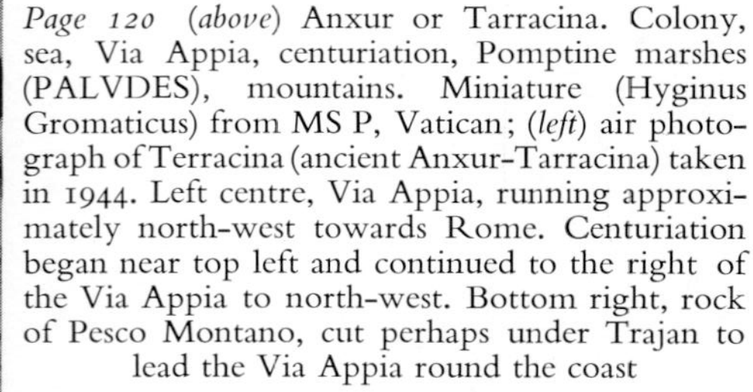

Page 120 (*above*) Anxur or Tarracina. Colony, sea, Via Appia, centuriation, Pomptine marshes (PALVDES), mountains. Miniature (Hyginus Gromaticus) from MS P, Vatican; (*left*) air photograph of Terracina (ancient Anxur-Tarracina) taken in 1944. Left centre, Via Appia, running approximately north-west towards Rome. Centuriation began near top left and continued to the right of the Via Appia to north-west. Bottom right, rock of Pesco Montano, cut perhaps under Trajan to lead the Via Appia round the coast

Asisium (Assisi), was intended. The Via Flaminia went past the walls of Hispellum. West of that road, coinciding with a *decumanus*, is the Foligno–Bevagna road (Fig 35), which has clear traces of crossroads

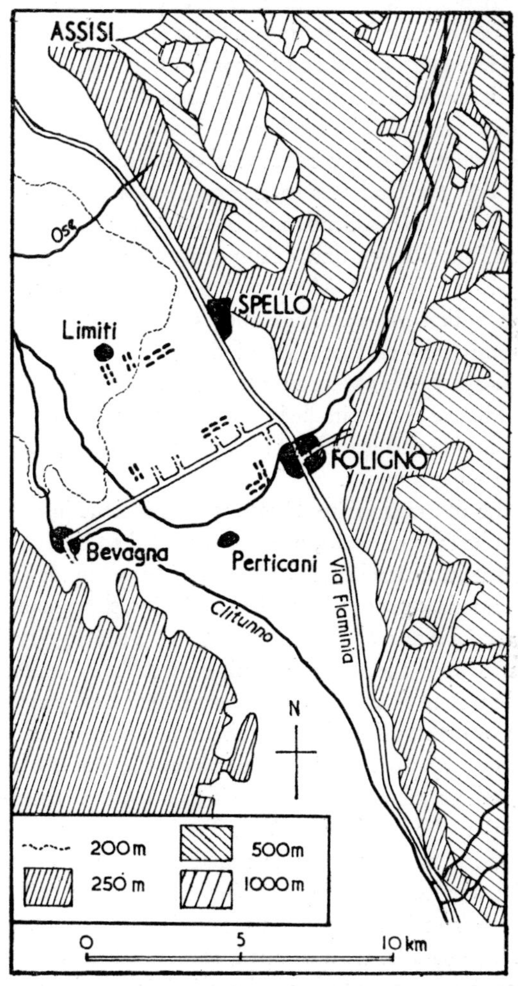

35. The area SW of Hispellum (modern Spello), showing remains of centuriation off the Bevagna-Foligno road. Note the place-names Perticani, from *pertica*, 'surveyor's rod' or 'centuriated area', and Limiti, from *limites*, 'balks', ie intersecting roads or paths

serving for centuriation at intervals of 720m or multiples. The miniature does not show two lakes, Umber and Clitorius, drained in the reign of Theodoric (AD 493–526) and in the sixteenth century respectively.

A colony depicted with centuriation round it and a mountain to the left (p 171) is shown by the inscription to be Suessa Aurunca (mod Sessa Aurunca). Here the text, that of Frontinus, reads: 'There are often farmlands, like those in the territory of Suessa in Campania, which have on Mt Maricus (see below) stretches of woodland defined by boundary marks. The ownership of these roads is claimed by legal process; for old maps show that the land was so assigned, as there was no assignable woodland adjacent to the farmlands.' Suessa Aurunca was first colonised in 313 BC and recolonised under Augustus. The hill near it was not as given, but Mt Massicus (M. Massico), famous for its wine. The division into squares which the miniature shows is at least in part incorrect, since Frontinus tells us that part of Suessa's land was divided into rectangles (narrow strips according to the *Liber coloniarum*). Strip cultivation would have been suitable in hilly areas where on partly volcanic soil the fertility varied greatly. The layout, however, cannot be confirmed from what remains.

One of the illustrations to Hyginus Gromaticus (p 68) gives us a colony with hexagonal walls, perhaps intended to be octagonal as the two sections are not attached. It is intended to explain the text: 'In some colonies founded later, as at Admedera in Africa, the *decumanus maximus* and *kardo maximus* start from the town and go through four gates . . . like camps. This is the most attractive method of centuriating: the colony embraces all four areas of the centuriated land and is convenient for cultivators on all sides, also the inhabitants have equal access to the forum from every part. Similarly in camps the *groma* is set up at the crossroads, where men can gather as if to a forum.' Ammaedara (the best spelling) is the modern Haïdra, Tunisia, refounded as a colony under Vespasian or his sons. The remains of the Roman settlement confirm the statement of Hyginus Gromaticus: the two intersecting Roman roads met in its centre, near which the Byzantine fort was later built. The parallel with camps is somewhat misleading, since the normal arrangement was to have the *principia*, headquarters building, occupying the centre of the camp.

In addition to these, there are a number of miniatures of colonies which are too vaguely described to be identified: a typical label is

Colonia Iulia. This title, if it refers to Italy, is likely to be a foundation of Octavian's;[16] if in the provinces, of Julius Caesar's or Octavian's. But it could stand for Colonia Iulia Augusta and represent a colony founded by Augustus after he became emperor; the latter, however, is more often simply called Colonia Augusta. In one miniature (p 172, top) this shorter title is given to a colony in a mountainous area, which is therefore likely to be Augusta Praetoria (Aosta), where Augustus settled veterans of the Praetorian Guard. The Roman remains at Aosta are extremely well preserved and keep the pattern of the ancient town-plan closely. But we cannot be certain of the identification, since the mountains round Aosta do not encircle it as in the miniature. The text speaks of centuriation being limited by the mountains. Aosta obviously had centuriation, but none has yet been found. A possible alternative is Augusta Bagiennorum (Bene-Vagienna).

In the case of Colonia Claudia (p 172), said to be near the territory of the Tegurini, there is no clear-cut identification. The Tegurini were a Helvetian tribe living near Aventicum (Avenches, Switzerland), which has a Roman amphitheatre and other notable remains. This was a colony and had centuriation, of which traces have been found,[17] but it was founded not by Claudius but by Vespasian under the name of Colonia Pia Flavia Constans Emerita. The R. Adum and Mt Larus, named on the diagram, are not known. Whether the criss-cross system of roads intersecting at the centre of the settlement represents a genuine tradition, as at Ammaedara, we do not know.

(b) *More than one settlement*. The largest maps preserved in the Corpus must be looked at with some suspicion. They are two found in the MS P, extending over double pages. These maps correspond to far simpler ones in A, which are different in appearance and have no wording. The first in P (p 171) shows a walled city in the centre, labelled Colonia Iulia Augusta and clearly intended to be Colonia Iulia Augusta Taurinorum (Turin). On the right is Hasta (Asti), 50km ESE of Turin. But here the accuracy ends. There is, as far as we know, no such place as Opulentia, to which two roads to the left lead. The closest corruption, Pollentia (Pollenza), is 50km SSE of Turin, not west; though the Peutinger Table, which spells it Polentia, has it SSW. Similarly there is some confusion over the phrase, corrupt in transmission, which must be intended to mean 'territory of the people of Segustero': presumably Segusio (Susa, Italy) rather than Segustero

(Sisteron, Alpes de Haute Provence) was in the mind of the map-maker.

But if this map is confused, the second in MS P is probably conflated arbitrarily from several areas. Atella is in Campania, Antemnae near Rome but reduced to a hamlet by Augustus' principate, and Vercellae (now Vercelli) between Milan and Turin; so that there can be no question of a single area of Italy. The multiple *kardines maximi* and *decumani maximi* point to there having originally been more than one diagram. So it seems quite likely that in origin these two types of map were one.

(c) *Plans of smaller areas.* These are sometimes concerned with centuriated land, sometimes with unsurveyed areas. The most conspicuous feature is a combination of plan and elevation; and regard for perspective[18] varies from diagram to diagram. The most glaring example of this is Fig 36. This is intended to illustrate the fact that the law required *limites* to be kept open as public thoroughfares, whereas some owners of hilly land inconsiderately stopped up the *limes* running

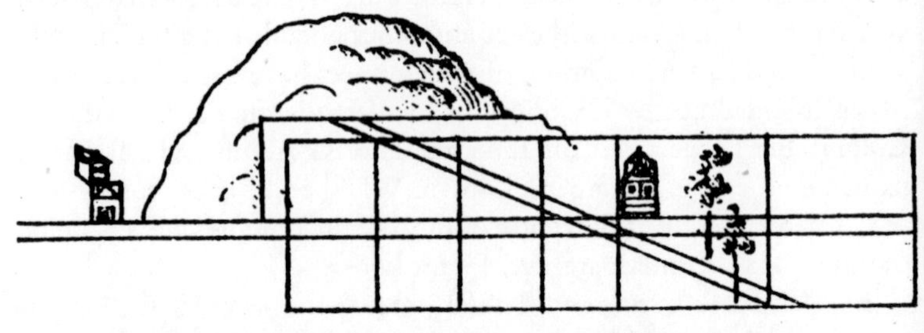

36. Diagram illustrating the keeping open of *limites*: Frontinus

through their property. The diagram is in two parts: the upper part shows a mountain in elevation, the lower part farmland in plan, with trees in elevation and country houses semi-oblique. There is a certain childishness and stylisation about some of the diagrams. Thus p 101 (below) is to illustrate the point about alluvial land quoted on p 107; here again part, the centuriated areas, is in plan, while the mountain and river, the latter with rather amusing ducks on it, are shown semi-oblique. Moreover the appearance of the diagram is not very different from that of p 172 (below), the purpose of which is to illustrate common pasture land. By placing this between two sets of squares the artist wishes to emphasise that it was not centuriated.

On another farm, lands are coupled with a mountain and an island. The text reads: 'A dispute about public ground belonging either to the Roman people or to colonies or boroughs arises whenever anyone has appropriated ground which has never been assigned or sold, eg an old river bed belonging to the Roman people, which the current, as there is an island in the way, has left dry, thus extending the adjacent private land; or forests, which in many places we know from old records belong to the Roman people; nearby, for example, in the Sabine territory on Mt Mutela. . . .' Although the name Mt Mutela is not known, there is no reason to suspect it as the name of a mountain in the Sabine hill country east of Rome. The farm, whose land also embraces the island, is named Septicianus after one Septicius, a common enough name in that area.[19]

It will be seen, in conclusion, that the maps and plans in the Corpus vary in their scope from very extensive areas to small-holdings, and that their value and veracity also vary, though not correspondingly. Despite their somewhat limited value, it is to be hoped that historians of cartography will in future devote more attention to this unusual collection of surveyors' teaching maps.

9

Roman surveying manuals

One of our chief sources of information about the Roman land surveyors is the Corpus Agrimensorum,[1] a collection of short works on Latin of very varying dates. It seems to have been compiled about the fourth century AD, and additions were made at various periods. Many of the treatises are illustrated, and most of the illustrations are very well preserved. It has been observed[2] that a great step forward in teaching techniques was made when manuscript manuals were interspersed with illustrations. This aid enabled pupils remote from the teacher in space and time to visualise objects, resulting in a dissemination of learning, and providing an ancestor for the closed-circuit television of today.

The earliest extracts are two short ones said to be from Mago and Vegoia, and one from the Lex Mamilia. Mago (p 34) was the Carthaginian writer on agriculture, and Vegoia or Begoe was an Etruscan nymph; but these are only attributions, and the date of the second, dealing with religious sanctions against violating boundaries, is the first century BC (p 229). The Lex Mamilia, from which sections 53–5 are preserved (p 106), was almost certainly part of Julius Caesar's legislation.

The manuals of earliest date, those by Frontinus, date from the late first century AD, but are truncated and mixed up with a commentary on them by Agennius Urbicus. The period following Trajan's campaigns on the Danube, AD 101–6, is perhaps the heyday of writings on surveying. But interest in the legal side of the subject continued throughout the Empire; collections were made of emperors' edicts, and extracts from relevant laws were compiled. As the Corpus grew, it came to be used for teaching purposes, so that it includes theoretical as well as practical instruction. Among the latest insertions, not found in our earliest manuscripts, are extracts from Boethius (ca AD 480–524) and Isidore (died AD 636).

In the Dark Ages and Middle Ages some Roman manuals perished, while others continued to be copied out. Among those which have not

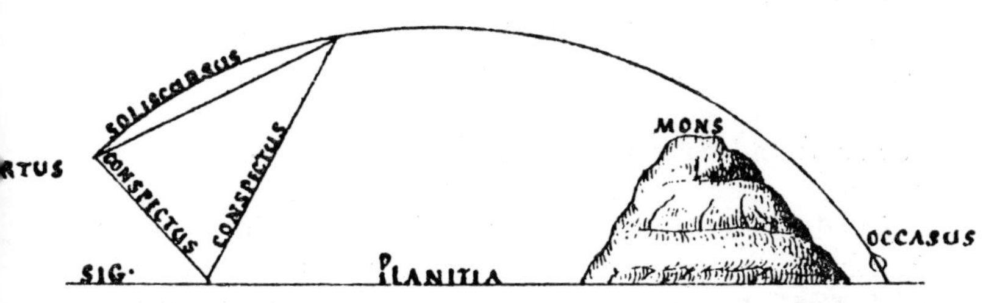

37. Miniature from MS A (Wolfenbüttel), the second oldest extant manuscript of the *Agrimensores*, to illustrate the fallacy of taking bearings by sunrise or sunset. 'If the KM or DM starts not far from a mountain, how can its course be properly surveyed, since the sun has set on the *groma* but is still shining on the other side of the mountain?'

survived, we must particularly regret the disappearance of manuals of road-making. A poem like that of Statius (p 41) or the modern excavation of a Roman road is no substitute for a first-hand account.

But although the monks preserved surveying material, their efforts have not resulted, in most cases, in surveying manuals in their original form being handed down to us. This is because (1) already before it reached them it had become corrupted and disordered, partly through being used for teaching purposes, (2) they did not understand the Latin, which is often technical and sometimes ungrammatical, and passed on their own errors and guesses, (3) the diagrams were copied by successive hands until they were far removed from their originals.

Nevertheless, we are fortunate in possessing several early manuscripts in uncial and minuscule hands, some with well-preserved miniatures in many colours and monochrome geometrical figures. Whereas in the Renaissance this manuscript evidence formed a large proportion of what could be gathered on the Roman land surveyors, today it has been supplemented by so much information from archaeology (including inscriptions discovered), map interpretation, and aerial photography that we can use it rather as one of several sources and check it whenever possible from the others. The patient labours of German, Swedish, and other scholars have plausibly sifted out texts from commentaries, suggested the correct order of paragraphs and restored corrupt texts. Comparative study of text and diagrams of the manuscripts has thrown light on their transmission, but more remains to be done.

The two oldest manuscripts of the Corpus are both bound in the same volume, which is in the Herzog August Bibliothek, Wolfenbüttel. They are of parchment, 31·8 × 24·8cm, and are known as Arcerianus A and B (catalogue number 36, 23 Aug 2°), the name coming from Johannes Arcerius, who owned the volume from 1566 to 1604.

Arcerianus A, occupying folios 2–83, in an uncial manuscript written in Italy and thought to date from the early sixth century. Folios 2–15 are in two columns, the remainder in long lines. It has illustrations in very good state of preservation, and in 1970 was published in black-and-white facsimile in the series *Codices Graeci et Latini*. The names of Adelbert and Giselbert which appear on folios 40 and 75 respectively are not those of the artist or artists of the miniatures, since they are tenth-century, whereas the miniatures are an integral part of the manuscript. We know that about 981 Abbot Gerbert, who later became Pope Silvester II, studied the works of the Agrimensores at the monastery of Bobbio in North Italy, being interested in all aspects of ancient science and mathematics, and having even studied in Spain the Arabic legacy from antiquity. Since in 1493 the Arcerianus was at Bobbio, it seems likely to have been the manuscript studied by Gerbert. An interesting attempt has been made to trace it back still further, though one must admit this cannot be proved.[3] Cassiodorus, whose interest in the Agrimensores has already been mentioned, retired from his active political life to the monastery of Vivarium in South Italy. There he built up a considerable library, which on his death he bequeathed to the monastery. It could well have included these two manuscripts, of which A at least was certainly written in Italy. The founder of the monastery of Bobbio was the Irish abbot, St Columban, who settled first in Switzerland and then in North Italy. Under his leadership the Irish monks who accompanied him wrote out many manuscripts, but we do not know whether Bobbio also acquired manuscripts from elsewhere. The later owners of the Arcerianus and dates of ownership are known;[4] among the owners was Erasmus from about 1527 to 1536.

Arcerianus B, occupying folios 84–122, is an uncial manuscript of the late fifth or early sixth century. It is inferior to A in most of its readings and has no illustrations.

The Palatinus (P, Palatinus Vaticanus Latinus 1564) was removed in 1623 from Heidelberg to Rome, since when it has been in the Vatican library. It is of parchment, 27·8 × 19·2cm, and dates from the ninth century. Its illustrations, like those of Arcerianus A, are in good state of preservation. An unusual feature is a set of four prefatory pages in vellum containing illustrations, to which titles were added by Metellus Sequanus in 1564. The first is a medallion with the bust of a young man, thought by Metellus Sequanus to be an emperor; he can hardly be Frontinus, as thought by Byvanck,[5] since that senator seems to have interested himself in technological writing not before the mature part of his career. The second and third show groups of surveyors seated in a circle deliberating, while the fourth shows a surveyor consulting an emperor, not this time youthful. Byvanck thought that the first originally belonged to a manuscript of about the sixth century AD, while the other three were painted by the same artist as the remainder of the manuscript. This is unlikely: the different material probably points to all four having come from an earlier manuscript and having been detached to prefix to the Palatinus. The other oddity is that a whole section of illustrations, the '*liber diazographus*', is grouped together, contrary to the usual practice of inserting miniatures at the appropriate point in the text.

It has been shown that the Palatinus is almost certainly the ancestor of a manuscript which, like the Arcerianus, is at Wolfenbüttel. This is the Gudianus, G, whose diagrams Blume[6] gives alongside those of Arcerianus A. Since there is little doubt that we possess the manuscript from which it was either directly or indirectly copied, those who wish to compare the diagrams of the two families of manuscripts should have recourse to Thulin's 1913 edition of the Corpus Agrimensorum. Thulin gives small black-and-white photographic reproductions of A and P where available or of the best substitute where either is missing. The newly published facsimile of A, mentioned above, should be an incentive to further research on the manuscript illustrations.

Manuscripts in Florence and Erfurt (Laurentianus Plut XXIX cod 32, ninth century; Amplonianus 362, eleventh century) contain only simple diagrams. A sixteenth-century manuscript in the Jena library is useful for supplying certain miniatures which have been cut out of A. A sixteenth-century manuscript in the Vatican, Vaticanus Latinus 3132, shows some independence in its diagrams. These and the other manuscripts vary considerably in order of treatises and in ascription of certain passages. None has a good clean text, some are fragmentary, and in some parts there is so much corruption that one may despair of reaching the right reading. Where figures are given, they sometimes differ between text and diagram, sometimes between one manuscript and another. Some extracts lack a beginning or an end. There has been much tampering with the text, partly by way of explanation, partly by users of the manuals who chose to alter the text for teaching or other purposes. Since, in addition to these defects, many of the writers of these manuals liked wrapping their instructions up in complicated phraseology, the task of the editor or translator is particularly difficult. Nevertheless, enough useful material can be extracted to make investigation well worth while.

The contents of the Corpus are listed in Appendix A. Of the treatises which provide a reasonable amount of information, the following may be mentioned. The excerpts from Frontinus are fragmentary and corruptly preserved, but authoritative. The commentary of Agennius Urbicus on Frontinus is late and not very helpful except for reconstituting the text. The work on measurements and geometrical shapes by Balbus gives us a fair specimen of the type of geometrical work which surveyors had to learn. There are two works ascribed to Hyginus, but they are evidently by different authors. The *De limitibus* ascribed to Hyginus is fragmentary but contains some useful information on centuriation. The *De limitibus constituendis* (or *Constitutio limitum*) ascribed to Hyginus Gromaticus is very helpful, though one has the impression that the author is to some extent groping in the dark, unable to understand some of the technicalities of land surveying which originated many centuries before he was writing. Siculus Flaccus, *De condicionibus agrorum*, deals with the legal status of types of land. The lists of colonies, with information about them, are discussed in Chapter 12. Nipsus writes on geometrical aspects of surveying. The *Casae litterarum* are lists of types of country estate under the late Empire expressed in terms of letters of the alphabet.

All of these are included in the *Schriften der römischen Feldmesser* by Blume and others. Most of them are to be found in Thulin's Teubner edition, which pays particular attention to variant readings in the MSS. As indicated, there is evidence throughout of interpolation, mutilation and padded explanations. There is no translation of more than short extracts; and indeed the Latinity and corruptions are a deterrent to the translation of more than selected passages.

The full set of illustrations appears in the Arcerianus A, in the Palatinus and its offshoot the Gudianus, and in some later MSS. One may speak of two families of diagrams, chiefly represented on the one hand by Arcerianus A, on the other hand by the Palatinus (the Gudianus may mostly be discounted). Of the two, the Arcerianus is on the whole the more accurate and sober. In some cases the two families have very similar diagrams, in others decidedly different ones. Certain MSS contain only geometrical drawings. I have attempted to classify the illustrations in *Imago Mundi*[7]: (a) surveying techniques in general, (b) surveying techniques: orientation, (c) illustrations relating to centuriation stones and boundary stones, (d) plans of towns and surrounding lands, (e) mapping, (f) illustrations of legal definitions, (g) theoretical instruction.

The colours are on the whole very well preserved. The simplest diagrams are monochrome, light brown. In the more elaborate illustrations we may see certain colour conventions observed. Roads are usually depicted red or brown, sometimes green. Water is blue or bluish-green. Buildings are mostly pale brown, yellow or grey; the predominant colour for roofs is red. Mountains are usually mauve or, if wooded, green; sometimes they are brown.

The idea of illustrating surveying textbooks with coloured drawings is not only a natural one but may be shown to have had a parallel in architectural manuals. Vitruvius[8] says of the architect: 'He must have skill in draughtsmanship, so that he may find it easier to represent by coloured drawings the effect he desires.' Does this refer to architects' plans and drawings or to textbook illustrations? We do not possess coloured illustrations of Vitruvius; but if he is thinking of the latter, then we may take it that many architectural and similar technical manuals were provided with them from the start.

It is difficult to say whether the earliest treatises in the Corpus were originally accompanied by illustrations or not. The texts of Frontinus and Balbus are greatly helped by these, but they never allude to them. Hyginus Gromaticus, on the other hand, uses the adverb *sic* to direct the reader's attention to a diagram, so that in the case of his treatise, unless someone did some careful editorial work on it later, the diagrams must have existed from the beginning. We do not know the date of this Hyginus (p 228), but his Latinity is respectable and therefore likely to be not too late. It seems probable that illustrations were first designed to accompany surveyors' manuals at a time when papyrus was still the regular writing material and the roll the usual form of book. The change from the papyrus roll to the parchment or vellum codex occurred in the third century AD. Parchment or vellum is an easier material for MS miniatures than papyrus, which even on the recto tends to have ridges. Perhaps some diagrams were first drawn on the one type and some on the other.

As the Corpus is specifically called a collection of the manuals of land surveyors, although it includes certain fringe topics it does not extend to subjects regarded as different specialisations. Many of these have perished, but among those surviving may be mentioned: (a) the section of Columella's agricultural manual dealing with mensuration; (b) the work of Frontinus on Rome's water-supply; (c) the treatise *De castrametatione*, 'On the Surveying of Camps'. This treatise, preserved except for its opening, is ascribed like two of the land surveyors' manuals to Hyginus. It contains a great quantity of detail about the internal subdivisions of military camps and the recommended measurements for each. Together with Polybius' account, it forms the basis of our 'literary' knowledge of the theory and practice of Roman military surveying. Likewise independent of the Corpus are (d) maps and itineraries, giving place-names and distances along main roads; (e) the Notitia Dignitatum, which gives lists of offices etc throughout the empire. Parallel to the Corpus, can other similar collections have disappeared? One can think of manuals of road-making, the building of bridges and aqueducts, building surveyors' compendia, works on the organisation of the army, fleet, and corn supply, and many other such treatises.

10

Centuriation

ORIGINS AND CAUSES

It is very probable that centuriation existed in Italy as early as the fourth century BC, and one may well look for a military origin of the practice. Colonies were thought of as agricultural settlements but at the same time as bulwarks of Rome's defence. If therefore their settlers were grouped together in convenient units, they would serve this military need better. The Roman mind tended to prefer squares to rectangles—Roman forts, with which the *groma* was also associated, are basically square. In Ennius and the grammarians we meet the phrase *Roma quadrata*;[1] whether this meant a square depression on the Palatine, in which the ceremonial plough and yoke connected with the foundation ritual were kept, or whether it denoted the square of the whole Palatine, it at least points to a hallowed association with a square area in the centre of early Rome.

As to *centuria*, that too had a military parallel, although its agricultural connections must not be forgotten. *Centuria* and *decuria* can denote tribal units of 100 and 10. Thus in Illyria[2] we hear of a boy aged ten who drowned, having been of the tribe Undii and in the second *centuria*. The smallest division in a legion, consisting theoretically of 100 men, was also known by that name. A 'century' must originally have been exactly 100 smallholdings, *heredia* ('inherited areas') of two *iugera* ('yoke areas') each. Although the terminology persisted, the number of holdings in a 'century' dwindled in time. Hence at Minturnae we find groups of 'centuries' associated with a cross-roads cult and based on common local interest.

It must be remembered that the usual Latin name for centuriation was *limitatio*. This stresses the importance of the *limites*, which could be anything from footpaths to wide roads. They served both for communication, especially outside the 'centuries', and for separating plots. Here the military parallel is with the *limes* of the whole Roman Empire, dividing it from the barbarians and serving as a line of communication for its frontier armies.

I. THE STUDY OF ROMAN SURVEYED LAND

WORK ON CENTURIATION

Remains of Roman centuriation have in some parts of the Roman Empire always been visible. In certain areas many squares are preserved in the form of roads or paths which have remained in use. Where these exactly or roughly correspond to Roman 'centuries', study of the remains would have been possible at any period. But whereas in the Renaissance Vitruvius was carefully read as a guide both to ancient remains and to trends in contemporary architecture, no Renaissance scholars applied themselves similarly to a study of the Agrimensores in order to understand the planning of the Italian countryside. Later, the picture in Italy was complicated by the existence of squares constructed for land improvement. Although metrology was well advanced, knowledge of the length of a standard Roman foot was long applied only to the study of buildings, while ancient land systems tended to be neglected.

It was in North Africa, as far as we know, that the first serious application of this knowledge was made. In 1833 a Danish naval captain, C. T. Falbe, noticed that squares round Carthage had sides of 708m. These he correctly equated with 2,400 Roman feet, forming a standard 'century' of 200 *iugera*. Since the area was known to have been colonised under Gaius Gracchus, there was never any doubt that the right conclusion had been reached.[3] Since then it has been discovered that this length of 708m is not absolutely uniform; nevertheless, the variations are only within a fairly narrow range.

In 1846 E. N. Legnazzi noticed that similar conditions applied in the Po valley. Two years later, at the same time as the first volume of *Die Schriften der römischen Feldmesser* by Blume and others appeared,

P. Kandler[4] studied the centuriation of the Trieste area. It soon became apparent that the square 'century' of 200 *iugera* was so generally adopted, particularly under the Empire, that any other shape or size must be regarded as an irregularity. A land square or rectangle which does not work out as having sides an integral number of *actus* long can only in fact be non-Roman; and caution is advisable before the discovery of an unorthodox system of centuriation is claimed. Yet it would be wrong to try to dismiss all these irregularities. Outsize 'centuries' are proved from Survey A of the Orange inscriptions, and are strongly suggested by observations in the field (p 85). The size 21 × 20 *actus*, reported in the Corpus as existing at Cremona, has been confirmed not only from there but from Aquinum; it results in a century of area 210 *iugera*. Clearly these irregularities were an acceptable alternative under the Roman Republic, when we also find allocations not always in integral numbers of *iugera* or unit fractions of a 'century'.

By the end of the nineteenth century, when Campania and other areas had also been investigated, A. Schulten was able to draw on a body of accumulated information about a number of centuriated areas. Improved maps of Italy and other countries led to greater precision in working out which modern roads or tracks corresponded to *limites* either external to, or within, 'centuries'. On the other hand traces of centuriation which must have persisted down to the nineteenth century became obliterated with the growth of cities. Those who work on the historical topography of Florence or Bologna, for example, may find that modern maps no longer reflect the Roman pattern, whereas recourse to eighteenth- or nineteenth-century maps may reveal an affinity. Greater interest in the visible remains led to more study of Roman surveying in the field. A significant discovery in Tunisia was one of centuriation stones near the Shott el Fedjadj, dating from Tiberius' principate.

During the early years of this century, W. Barthel worked extensively on centuriation in North Africa, principally Tunisia. Areas of France, Germany, and Switzerland were investigated, though the evidence was not always certain. Many scholars worked on centuriation in Italy; the most famous was P. Fraccaro, who made numerous contributions to its study, and the exhibition of Roman surveying which he set up may be seen in the Museo della Civiltà Romana (formerly Roman Empire museum) in the EUR area near Rome. The displays in this and related subjects are well worth visiting.

In the two World Wars the technique of aerial photography and its interpretation were developed. Pioneers in its archaeological use were Léon Rey (Macedonia, 1917), T. Wiegand (Palestine, 1919), Beazley (Mesopotamia, 1919) and Cecil Curwen (England, 1929). After World War II large quantities of aerial photographs became available for study, and more were specially taken in areas likely to produce results. Different types of study emerged from different areas. In the Sahara the Roman *limes* (frontier of the empire) was plotted over vast areas. One of the pioneers in the study of centuriation from aerial photography was John Bradford, who studied it in Italy, Yugoslavia, Tunisia, and France. In some areas systems of different dates running at divergent angles could be seen superimposed. On the Dalmatian coast the *kardines* could be observed continuing along the same line after being broken by a large gap of water. In the case of Valentia, the modern Valence (Drôme), Bradford showed that the *limites* needed for modern communications persisted, whereas many of those at right angles to them tended to disappear. In Britain, air survey has been particularly valuable in discovering new marching camps and filling in hitherto unknown areas of Roman activity; most helpful work on this has been done by Dr J. K. St Joseph.

Aerial photography and ground survey have been used to complement each other. Details which cannot be seen on the surface can be plotted by air photographs and checked by excavation. Sometimes the existence of a *limes* between or within 'centuries' can only be demonstrated if the photograph has been taken with a low sun at a suitable time of the year, so that differences in crop markings appear. In Tunisia C. Saumagne discovered as early as 1929, from aerial photography, the remains of centuriation between El Djem and the sea. Areas of Italy whose ancient land divisions were first spotted from the air include Alba Fucens, Cosa, and Luceria. In Apulia the connection of centuriation with Roman agriculture has been studied by British archaeologists, using aerial and ground survey. Among the places for which additional information has been obtained from air photography are Florence, Pisa, Cesena and Orange. By results obtained from both ground and air, it has become possible to prepare atlases of centuriation. The *Atlas des centuriations romaines de Tunisie*, which appeared in 1954, covered a wide expanse of that country, using over 15,000 negatives taken between 1947 and 1951. More recently areas of North Italy have been covered by two volumes of the *Tabula Imperii Romani*, namely Mediolanum (Milan) and Tergeste (Trieste).

Some research, too, has been carried out on the study of place-names in centuriated areas. The etymologies involved are normally of three types: (a) from surveying terms; (b) from distances; (c) from Roman names. There are also many names derived from types of produce, natural features of settled landscape, etc. For this type of investigation research into medieval records is indispensable, as quite often there has been a change in place-names since the Middle Ages, or the medieval spelling may give a clue to the true derivation.

The most interesting survival in the first category is the word *pertica*. It occurs in a place-name near Padua, S. Giorgio delle Pertiche. This is in an area where the regular network of centuriation has left one of its strongest imprints. The word *pertica* in Italian and Spanish means a surveyor's rod, and *perticatore* is an Italian word for a land surveyor. In French it became *perche*, whence English 'perch'; and this, as in Latin and certain modern languages, can denote a measure. *Limes* and its derivative *limitatio*, the regular Latin for centuriation, have left their derivatives in Romance languages. In parts of Italy *limes* appears also in place-names, eg from the Po valley, Limidalto, Limisano, Solimite.[5] Similarly *fines* appears in French as Fins.[6] *Cardo* led to Cardito in Campania and to Cardeti in the Po valley; *centuria* is also thought to have survived in place-names. As to *decumanus*, it has been shown by Campana[7] that we have to be careful. There was a place in the Padua area called Desman, which may have been derived from it. There is, however, also a straight road from Ravenna to Cesena called the Dismano, which coupled with the fact that in 1172 it was called *via decimani* shows that it is derived from *decumanus* (*decimanus*) in some sense. But apart from doubt whether it should not be a *kardo* rather than a *decumanus*, it is known that in the Middle Ages the whole area was called Dismano, and this goes back to a Latin phrase of AD 896 *territorio Ravennate in Decimo*. So it may perhaps mean the tenth mile from Forum Popili.

This leads to place-names derived from numerals. Here we find many names, especially in Italy, some from cardinal, some from ordinal numbers. Fossa de Quarto (Po valley, AD 1108) may be the ditch at the fourth *kardo* or *decumanus*. Quinzano in the same region probably comes from *quintana*, one of the streets in a military camp. Many place-names with a numeral component, on the other hand, come from distances in miles. Tricesimo comes from *Ad tricesimum*, 'at the thirtieth mile' from Aquileia; Sesto Imolese and others could have a similar derivation. The Spanish *quinta*, 'farm', comes not from a *limes* or from a distance but from a one-fifth farm levy. On the other hand the Spanish place-name Quintanilla may be a diminutive from *quintana* as above.

Finally there is the very large group of French and Italian place-names derived from farms which in turn were derived from property-owners. The usual adjective formed in classical times from *nomina* was in *-anus*, eg Septicius gave the adjective Septicianus, and his farm was so described. These forms in course of time often became village or town names, eg in the Po valley Bavignano from Papiniano, 'farm of Papinius', Fiagnano from Flaminiano, 'farm of Flaminius'. Although this suffix is found in France, eg Perpignan from Perpennianum, an originally Gallic suffix *-acum* also appears, giving rise to such names as Cognac, well known to the wine trade, from Cominiacum, Jonzac from Juventiacum, etc. Some road names derived from their planners survived, and by this means Aemilius left his name to Emilia.

Various aspects of centuriation are currently being investigated; thus, F. Castagnoli is particularly interested in the connection between Roman surveying and Roman towns and town-planning; General Giulio Schmiedt in the land systems of the Greek colonies in southern Italy; R. Chevallier in the largest areas, Tunisia and the Po valley; G. D. B. Jones in the connection of centuriation with Apulian agriculture.

The areas where centuriation is known are outlined in part II of this chapter. The largest are still, and will certainly remain, Italy and Tunisia; but France, Germany, Switzerland, and Yugoslavia are also of some importance. Britain has so far yielded only one likely example, near Rochester (p 191). Areas where more centuriation may well be discovered are in Spain and Asia Minor.

UNORTHODOX SYSTEMS

The enquirer into Roman centuriation should bear in mind (a) that in some areas the land has been divided up in more recent times into squares that have no connection with Roman centuriation, (b) that side by side with Roman centuriation other forms of land division or usage existed.

The idea of dividing agricultural land into squares was not an exclusively Roman one. We have seen that Herodotus thought all the land in Egypt was divided into squares, and although this was not so, some certainly was. In Japan in the seventh century AD[8] towns known as the Kofuku had a regular network of squares on the alluvial plain, the lines serving as roads and drainage channels. The orientation was by the four compass points, and the system was known as Jóri, Jo signifying the north–south lines and Ri the east–west. As opposed to the Roman squares of 705–10m, the Japanese squares had sides of only about 120m. The system was devised at a time of social reform and was operated by a rigid bureaucracy.

The eighteenth century was in many parts of Europe a period of land improvement. Some of the improvement systems, particularly in Italy (*bonificazioni*), took the form of division of the land into squares, and these may deceive the researcher. Where there is doubt whether a system goes back to Roman times, (a) one should establish whether the *limites* are a definite number of *actus* apart, preferably the regular number 20 × 20, but in any case not a prime number of *actus*; (b) any possible local references (eg place-names indicative of surveying) should be investigated; (c) early maps, preceding any likely improvement systems if possible, should be examined; (d) use should be made of aerial photography as well as ground survey.

Centuriation or allocation by parallel lines only, and systems of *strigae* and *scamna*, have already been mentioned. But it must be remembered that in many areas of the Roman Empire land was not centuriated. There the pattern of land division varied according to local circumstances, dictated by history, agriculture, climate, and terrain.

From the historical point of view, the majority of towns which were not colonies and had no *ager publicus* nearby chose not to have their land centuriated. If an efficient but irregular system operated well enough, they would tend to preserve it. In Roman Britain there were small corn-plots of the same type as the earlier lyncheted fields. Some farms continued to be worked under the Roman occupation on the same basis as before.[9] Similar corn-plots are to be found in parts of Germany and Holland. A good example from the north of the province of Limburg is shown in Fig 38. Many intersecting lines, north-east/south-west, may be seen, but no overall pattern in terms of *actus* is discernible. Country estates in Germany, such as one near Mayen,[10]

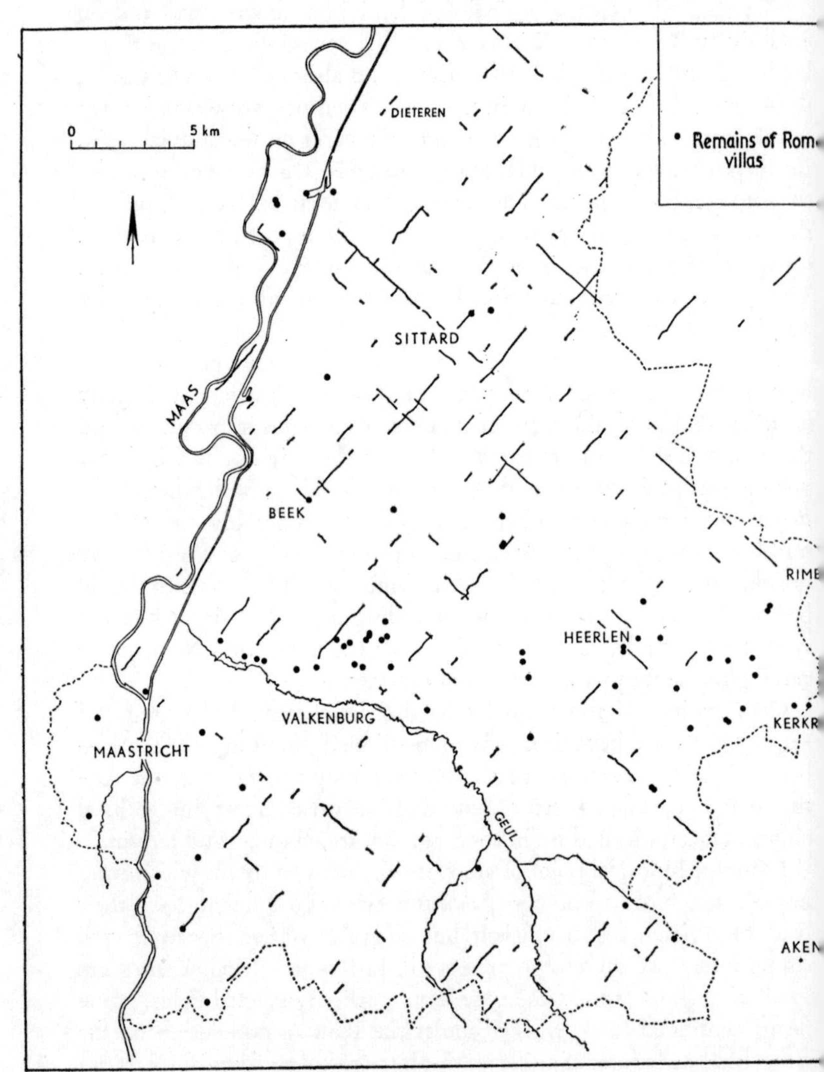

38. Remains of ancient land division in Limburg province, Holland

show parcelling of land in strips connected by paths at right angles, to some extent following the contours. In the Vosges, the hillside is contoured with terraces called Rotteln, some of which date from the Roman period.[11] In the Mt Hymettus district of Attica, levelled field-banks and terracing have been studied both on the ground and from air photographs.[12] In Southern Algeria, where farming was practised despite the shortage of water, small fields were clustered round irrigation points. Other such irregular schemes could be adduced from different parts of the Roman Empire.

The estates attached to country houses and farms show a wide variety of field pattern even within the same province. Excavation and aerial photography have revealed the plans of a number of these. It is clear that although owners of estates called in surveyors for various purposes, they did not normally plan their estates on even a modified system of centuriation. Thus, in the land between the villa and the R. Aire at Gargrave, Yorkshire, which is at present being examined, what appear to be drainage channels of the late Empire radiate outwards from the villa rather than parallel to each other. Research on the field systems of Roman villas is still not fully developed, but it is more advanced in Britain than in many areas and it may make a substantial contribution to our knowledge of Roman land use.

As has been mentioned, the planning of towns was clearly not the sphere of the land surveyor. Measurements were made in feet, not in *actus*; there was no centuriation; streets were planned sometimes parallel and at right angles, sometimes less regularly; architects, not surveyors, were responsible for erecting and maintaining buildings. The result is that only where a new settlement is planned from the start on empty land does a pattern common to town and country emerge. In other cases the only feature likely to be shared is a *decumanus maximus* and a *kardo maximus*; and even these have been assumed for some towns where the evidence must be considered doubtful.

II. CENTURIATED AREAS

A great number of colonies, as well as certain non-colonies, had centuriated land, and quite a large proportion of this land has left its mark on the landscape. But the state of preservation of such marks varies enormously, as does the extent of the lands. The two largest areas are in northern Italy and in Tunisia. In parts of the Po valley mile upon mile of modern roads and farm tracks represents exactly or approximately the line of the ancient network. In Tunisia, where observation in the field goes back nearly 140 years, a fair proportion of the gigantic extent of centuriation, where it has not been swallowed by the sand, has been made out with the aid of aerial photography.

Certain areas of Italy are of particular interest to the historian because they are comparatively well documented. Apulia shows the correlation of centuriation with agriculture, while Campania is a settled and well-populated area which in places reflects continuity from ancient times. In southern France the remains of what were extensive systems are comparatively scanty, so that correlating the inscriptions of Orange with what survives in the Rhône valley today is a difficult jigsaw puzzle. In other provinces the state of preservation varies. Thus parts of the Dalmatian coast have comparatively well-preserved centuriation, whereas in Spain, where we know from the Corpus that it existed, it is not known to have left any trace, though detailed air photography might well reveal patterns.

The summaries which follow do not attempt to analyse particular areas in depth. The literature on this subject is enormous, and it is only hoped, by pointing to some of the main features, that those who are interested in individual regions will be encouraged to make a detailed field study of these, and that the general reader will turn to fuller accounts. The great majority of the systems listed are of standard 20 × 20 *actus* squares; this should be taken to be the case if there is no indication to the contrary. In the Bibliography (pp 242–7), local sections are shown separately, but only a brief selection has been made of that on Italy, as explained on p 242.

ITALY

Latium and Campania, the homelands of Rome's early expansion, contain, as may be expected, some of the oldest centuriation (Fig 39). A good example on a small scale is that of Anxur-Tarracina (Terracina), described on p 116. Despite its smallness, the area is of particular interest because we have details of the foundation of the colony, because it is mentioned in the Corpus and because the local historical

39. Map of sites in Italy where centuriation is known to have existed

geography merits particular attention. At Aquinum (Aquino) there are said by Castagnoli[13] to be two systems, (a) in almost square rectangles, 1250 × 1275m, (b) farther west, in rectangles of 21 × 20 *actus*. As to the first of these, whereas the longer sides are close enough to 36 *actus*, 1277·4m, the shorter ones are too long for 35 *actus*, so that one may be doubtful if this can be considered part of a genuine centuriation system.

In the plain of Capua, *ager Campanus* (p 207), we find one of the best preserved examples: the whole area between the ancient Capua (S. Maria di Capua Vetere) and Marcianise has very clear signs of centuriation. It is a fertile region, which produced large cereal and vegetable crops. The Etruscans established themselves in Capua about 600 BC, but it was taken over by Sabelli about 440. During the Second Punic War it sided with Hannibal, and on being recovered by Rome in 211 BC was severely punished. The territory of Capua became *ager publicus*, mostly rented out; but in 59 BC Julius Caesar settled 20,000 colonists on it. A centuriation stone,[14] bearing the names of a Gracchan land commission of three, in the original spelling C. Sempronius Graccus, Ap. Claudius Polcer, P. Licinius Crassus, and the 'century' number SDI KKXI, was found detached near S. Angelo in Formis. It has been reckoned that, if it was in fact *in situ*, the probable central intersection was at what is still a crossroads.

At Minturnae (near Minturno), mentioned in the Corpus, there are only faint traces south-east of the colony, but a large number of cross-road altars, *arae compitales*, have been found.[15] The twenty-nine stones found in 1932–3 each contain twelve names of freedmen, freedwomen and/or slaves, often divided into four groups of three, a number of them stating that the named *magistri* (*magistrae*), 'masters (mistresses) of crossroads', dedicated the altar to a particular god or goddess. Staedler, who dates them to 28 BC, explains each group as representing a 'century'. Every four 'centuries', grouped round a *compitum*, 'crossroads', constituted a *pagus* which duly set up an altar. Thus if there were originally 30 *pagi*, there must have been 120 'centuries'. At Cales, south-east of Calvi Risorto, there is centuriation by *decumani* alone.[16] There are traces of centuriation at Allifae, Nola and Nuceria (Alife, Nola, Nocera).

In Apulia there is centuriation at Aecae (Troia), partly on the line of the Via Traiana, with an extension to the north; at Luceria (Lucera),[17] where the early allotments were either by *decumani* alone or in rect-

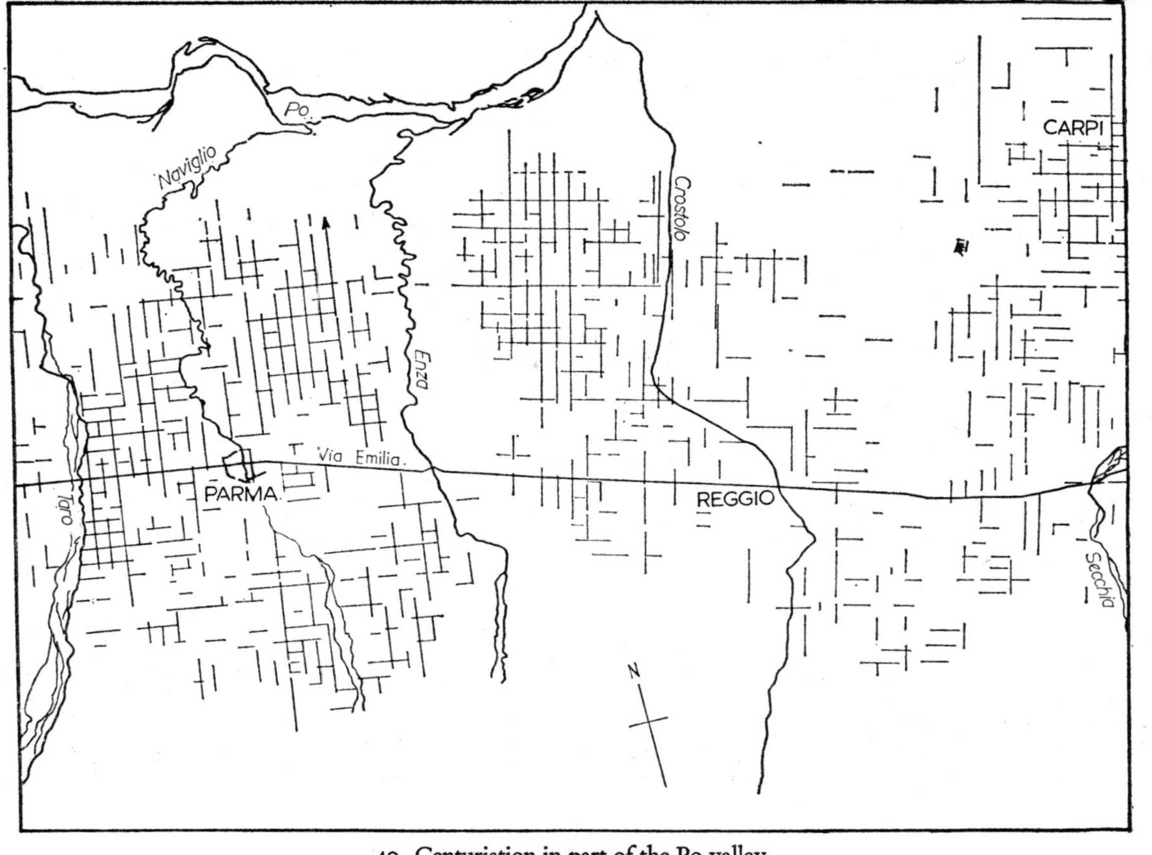

40. Centuriation in part of the Po valley

angles, though there is also a small area of regular squares; at Herdoniae (Ordona), where there are two sets of centuriated land, to north and 5–6km south; at Asculum (Ascoli Satriano), where there are two sets superimposed; and at Ergitium, east of S. Severo, where there appear to be rectangles of 20 × 16 *actus*. A centuriation stone has been found near Luceria, and near Cannae the stone mentioned on p 108. Roman Apulia has been studied, among others, by John Bradford and G. D. B. Jones, and the latter is at present writing a book on it.

In Lucania, evidence from a Gracchan centuriation stone shows that there was division of land in the territory of Atina, in the Vallo di Diano. In Samnium the centuriation of Alba Fucens, in the plain of Avezzano, where *decumani* alone were earlier found, has been shown also to contain rectangles in the central area and squares on the hill.[18] In Umbria the centuriation of Fanum Fortunae (Fano) is visible, and in the territory of Hispellum (Spello), mentioned in the Corpus, clear traces emerge as side-turnings off the Bevagna–Foligno road, which preserves the line of an ancient road (Fig. 35).

In Etruria the centuriation of Cosa emerged from air photographs as by *decumani* alone. The streets of the ancient town are not aligned with these, but the Capitolium is.[19] Florentia (Florence) had a very large expanse of centuriated land; the difficulty in studying this is that modern development has largely removed the traces of ancient roads. Other examples in Etruria are east of Pisae (Pisa), in the plain of Luca (Lucca), and in the territory of Luna; this is the modern Luni, but the land concerned is near Pietrasanta.

The best region anywhere in the Roman Empire to see very extensive centuriation well preserved is the Po valley (Figs 40 and 41). Much work on this was done by the late Plinio Fraccaro, and in more recent times a host of researchers, especially G. A. Mansuelli and R. Chevallier, have been engaged on it. Whereas in other parts of Italy the centuriation is mostly disconnected, here it is in many places continuous. It is aligned with the Via Aemilia, was connected with the great colonising effort of the early second century BC, and comprised all but what were marshy areas to the east. The Via Aemilia, running 280km north-west and WNW from Ariminum (Rimini) to Placentia (Piacenza), was built by Marcus Aemilius Lepidus in 187 BC. It connected many of the important towns of the Po valley, and has given its name to the modern Emilia. The regular centuriation, 20 × 20 *actus*,

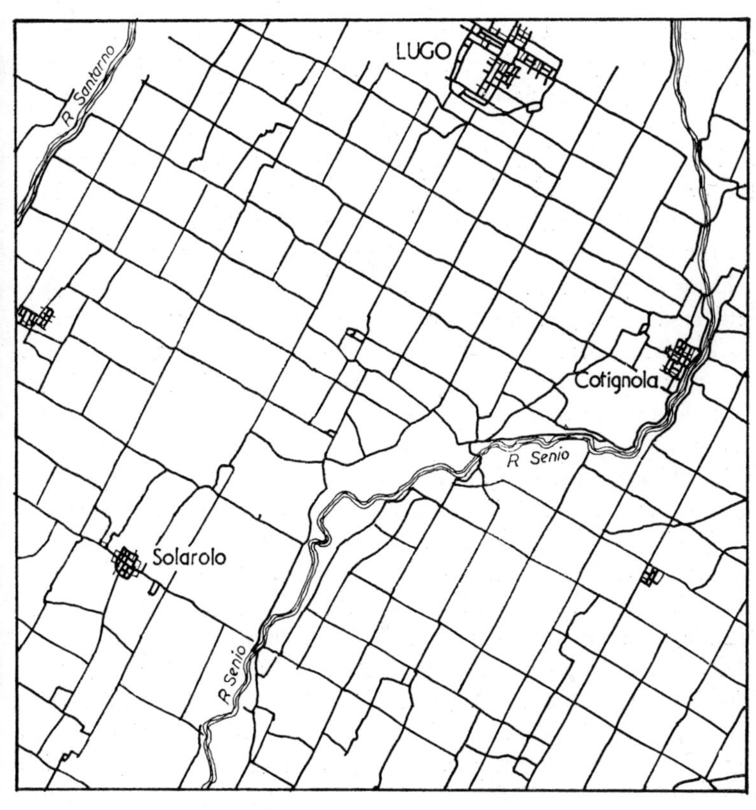

41. Modern roads near Lugo, Po valley, which preserve the pattern of centuriation

follows fairly closely the successive orientations of the Via Aemilia. It is therefore likely to be part of a single plan, amplified from time to time, and to date in its original form from not long after the building of that road. One suggested date[20] is 173 BC, when Livy[21] says Romans were given 10 *iugera* a head, allies three, of land in Cisalpine Gaul. But Mutina (Modena) and Parma were founded ten years earlier, so it may perhaps be hazarded that 183 BC was the starting date.

The centuriation schemes of Ariminum and Caesena (Cesena) are differently orientated, as are one of the two systems round Forum Popili (Forlimpopoli) and a grid, perhaps of later date, north of

Bagnacavallo.[22] In the western zone, between Placentia and Parma, there are three orientations, two corresponding to lines taken by the Via Aemilia, the third to its continuation westwards from Placentia.

In some cases the territory allotted to these colonies was enormous: an inscription from Clastidium[23] shows that this town was served by the *pertica* of Placentia, although the two are 49km apart. In each case where the original allocation and the number of first settlers is known, the total extent of centuriated land is far greater. Thus at Mutina we are told by Livy[24] that at the foundation of the colony there were 2,000 settlers who were given 5 *iugera* each. This makes 10,000 *iugera* or 50 'centuries', compared with 500 'centuries' traceable on the ground. In other words, there must have been successive allocations, each extending the grid further, probably up to and including a settlement under Augustus. The same applies to Parma, with 2,000 settlers given 8 *iugera* each, making 80 'centuries' out of 900 visible. The biggest holdings were at Bononia (Bologna), 50 or 70 *iugera* for each settler.

Liguria was a hilly area not very suitable for centuriation, but there are remains at Dertona and Pollentia (Tortona, Pollenzo).

In Venetia centuriation is preserved at a large number of places: Vicetia, Patavium, Acelum, Tarvisium, Altinum, Opitergium, Feltria, Belunum, Iulia Concordia, and Aquileia (Vicenza, Padua, Asolo, Treviso, Altino, Oderzo, Feltre, Belluno, Concordia, Aquileia). They are regular in size except at Acelum and Tarvisium, both of which apparently have 'centuries' of 21 × 21 *actus*, and Iulia Concordia, which may have had 'centuries' of 20 × 18 *actus*.[25] Aquileia, now well inland, was a large harbour city under the Empire, and its centuriation extended a long way inland. It may have included Utina and Forum Iuli (Udine, Cividale), while ad Tricesimum (Tricesimo) may have belonged to Iulium Carnicum (Zuglio). On the other hand, Forum Iuli and ad Tricesimum both have irregular 'centuries', perhaps 12 × 12 or 23 × 12 *actus*, so that they may have the same origin. Perhaps all, including Iulium Carnicum, were subordinate to Aquileia.[26]

Istria became part of Italy under Augustus, and is therefore included in this section. It has centuriation at Tergeste, Parentium and Pola (Trieste, Poreč, Pula), the last of which has a well-preserved system. An inscribed stone at Parentium mentions a main road 20ft wide, *via publica lata pedes XX*.

In Transpadane Gaul, centuriation is preserved at Cremona, with 'centuries' of 21 × 20 *actus*, Brixia (Brescia), Mantua, Verona, Eporedia (Ivrea), Augusta Taurinorum (Turin), Ticinum (Pavia), Laus Pompeia (Lodi), in the Oglio valley, where there are conspicuous parallel roads, and in the plain from there to Mediolanum, Novaria, and Vercellae (Milan, Novara, Vercelli). In the Cottian Alps there are remains of land division near Caburrum (Cavour).

GAUL AND ADJACENT AREAS

France

One of the best preserved centuriation sites in France is Valentia (Valence-sur-Rhône). As one drives out of the town towards Chabeuil, one can see how the east–west *decumani* survived because they linked these two places, whereas the north–south *kardines* had less useful function, so often disappeared.[27] In the Orange–Avignon area there seem to be about five different schemes, presumably all centred on Arausio, superimposed at different angles[28] (Fig 45). The one immediately south of Orange is rectangular, the others square. Tracing the patterns in the field is difficult, and there is doubt about their exact correspondence with the three cadasters found on the inscribed stones and described in Chapter 11. The centuriation of Narbo (Narbonne), capital of Gallia Narbonensis, has been traced from the air.[29] Adjacent to this but on a different orientation is that of Baeterrae (Béziers), which was also a colony. Another area has been traced between Forum Domitii and Sextantio (near Montpellier). The centuriation near Arelate (Arles) is very broken: evidently an unsuccessful attempt was made to centuriate and allocate the fluvio-glacial Palines de la Crau.[30] In other areas of France, eg Normandy and Brittany, round Rennes and in various parts of Burgundy, the evidence is very uncertain.

Britain

Details of the somewhat meagre discoveries are given in Chapter 13.

The Low Countries

Research has been carried out in Belgium and Holland since World War II on areas where there are land divisions in straight lines and at right angles. Fig 38 shows villas and presumed Roman *limites* in the south of the province of Limburg (Netherlands). The lines indicated on the diagram, running at 42°–222° and 132°–312°, ie approximately north-west/south-east and north-east/south-west, are roads, but a detailed plan of the Sittard area[31] shows also municipal and field boundaries. In some cases these lines coincide with regular divisions of 20 *actus*, in other cases they do not. Fainter traces of this same orientation may be seen throughout the province.

In Belgium, the principal remains are (a) between Maastricht and Tongeren, North Limburg, squares with a single orientation; (b) south-west and west of Tongeren, squares and rectangles with various orientations.

Germany

None of the areas suggested in Germany is conclusive of centuriation. Near Cologne there are thought to have been squares of 1,600 × 1,600ft, joined in *saltus* of 4,800ft. One may observe (a) that Siculus Flaccus speaks of *saltus* as a square measure (p 191 below); (b) that although 4,800ft makes 40 *actus*, 1,600ft is not an integral number of *actus*; (c) that some centuriation lines suggested by Klinkenberg[32] for that area are likely to be unsound, since Roman buildings go right across alleged *limites*.[33] There may, however, be some sort of non-standard divisions, such as have been found in Rhaetia and Noricum. An inscription from Cologne[34] speaks of a *scamnum*, ie rectangular holding. Other remains are reported from Kreuznach, Pfeddersheim, Alzey, near Mainz, Friedberg and Wetterau. Whether an abbreviation in an inscription from Germany[35] stands for four 'centuries' is uncertain.

Switzerland

Three areas have yielded results: Noviodunum, Augusta Rauricorum and Aventicum (Nyon, Augst and Avenches), all three colonies. Avenches, as mentioned on p 123, may be portrayed in a miniature of the Corpus. As to the Augst area, Déléage[36] thought that its land divisions might well be those of a private estate.

DALMATIA

Most of the centuriation follows the north-west/south-east line of the Adriatic coast. There are conspicuous remains at (a) Iader (Zadar), together with the island of Ugljan, perhaps ancient Lissa,[37] colonised by Augustus; (b) Salonae, near Split, and Tragurium (Trogir). Although the centuriation belonged to Salonae, the palace to which Diocletian retired in AD 305 had its Golden Gate on a *decumanus* leading to Salonae. The system extends over the bay on the same orientation. To achieve this, either a system of high marks could have been used in plotting the original survey or measuring could have been carried out round the coastal indentation. The land immediately round Salonae does not seem to have been centuriated; (c) the island of Pharos or Pharia (Hvar) has NNE–SSW centuriation, thought by Bradford to be in squares of 5 × 5 *actus*.

Suič[38] has reckoned the area and settlers of the Istrian and Dalmatian colonies as follows, arbitrarily allowing 50 and 66⅔ *iugera* for each settler (head of family):

Colony	*'Centuries'*	*iugera*	*Colonists*
Parentium	450	90,000	1,640
Pola	650	130,000	2,369
Iader	50	10,000	173
Salonae	80	16,000	260
Epidaurum	50	10,000	173

Aerial survey has shown that all the available arable land was taken up. But if, for example, the figure of 173 heads of family as *coloni* is correct, only about 1,000 in all can have been settled on the centuriated area, whereas the urban inhabitants were quite numerous: the amphitheatre accommodated 14,000 or 15,000, and the water supply at the end of the first century AD may even have sufficed for 40,000.

GREECE

The Romans left most of Greece unchanged in organisation, so that we should not expect to find much centuriation. The only example which can be clearly seen from the air is Nicopolis,[39] where the *limites* are closely connected with the streets of the town. At Pella and Thessalonica only faint traces were discernible from air survey. The absence of any sign of centuriation at Corinth may mean that in 111 BC its land was only measured and bounded and not centuriated, and that, as scholars think likely, Julius Caesar's foundation was not an agrarian colony.

NORTH AFRICA

There are vast centuriation schemes in Tunisia, but little elsewhere (Fig 42). The former have been carefully mapped on a scale of 1:50,000 in the *Atlas des centuriations romaines de Tunisie*.[40] The following are the principal areas:

(a) An area bounded by the sea at Hippo Diarrytus (Bizerta), by the

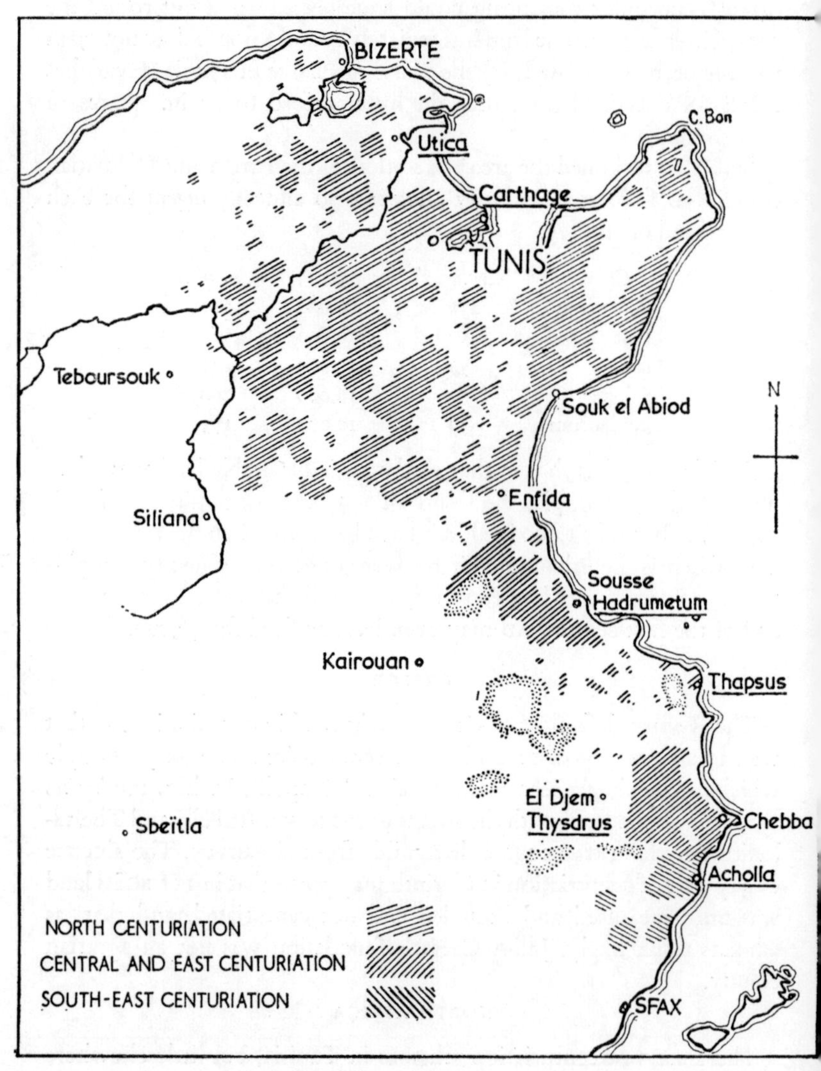

42. Map of centuriation sites in Tunisia

153 (*above*) Minturnae. Colony, sea, River Liris, mountains, bronze statue (AENA), new
cation of lands (ASSIGNATIO NOVA). Miniature (Hyginus Gromaticus) from MS P, Vatican;
(*below*) Minturnae, near modern Minturno, the forum area

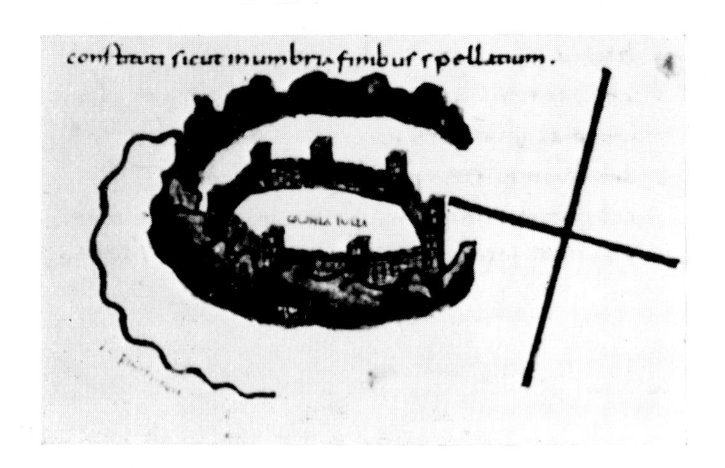

Page 154 (*above*) Hispellum (modern Spello; already *finibus Spellatium* in the text). Col (COLONIA IVLIA), river acting as boundary of territory (FL. FINITIMVM), centuriat Miniature (Hyginus Gromaticus) from MS P, Vatican; (*below*) centuriation planned round an exis settlement. There should be only one KM, indicated by the wider horizontal line. Minia (Hyginus Gromaticus) from MS A, Wolfenbüttel

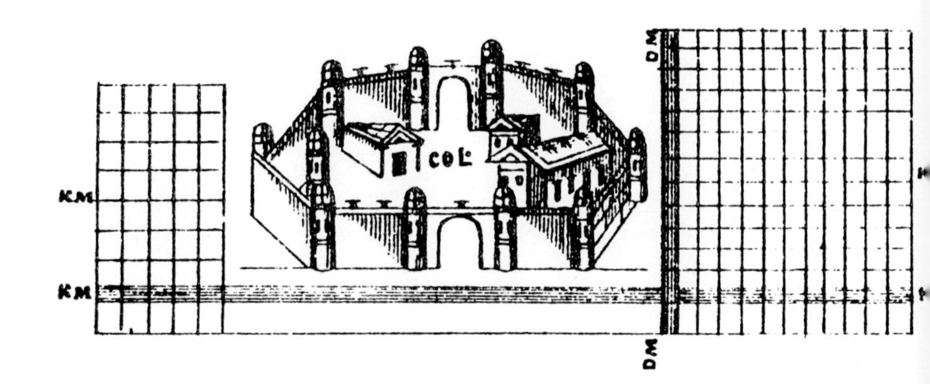

modern Enfida and by the Wadi Siliana. This corresponds to the Gracchan allocation, and faces sunrise at the summer solstice.

(b) A faintly preserved network in the valley of the Mejerda.

(c) An area north of Hadrumetum (Sousse).

(d) An area round Acholla, facing sunrise at the winter solstice.

(e) The centuriation schemes of Ammaedara and Sufetula (Haïdra, Sbeitla), which presumably tied up with the scheme of AD 29 mentioned below.

Many areas have a single continuous network, but around Enfida about five different orientations are visible, based on squares and rectangles of various sizes (Fig 43).

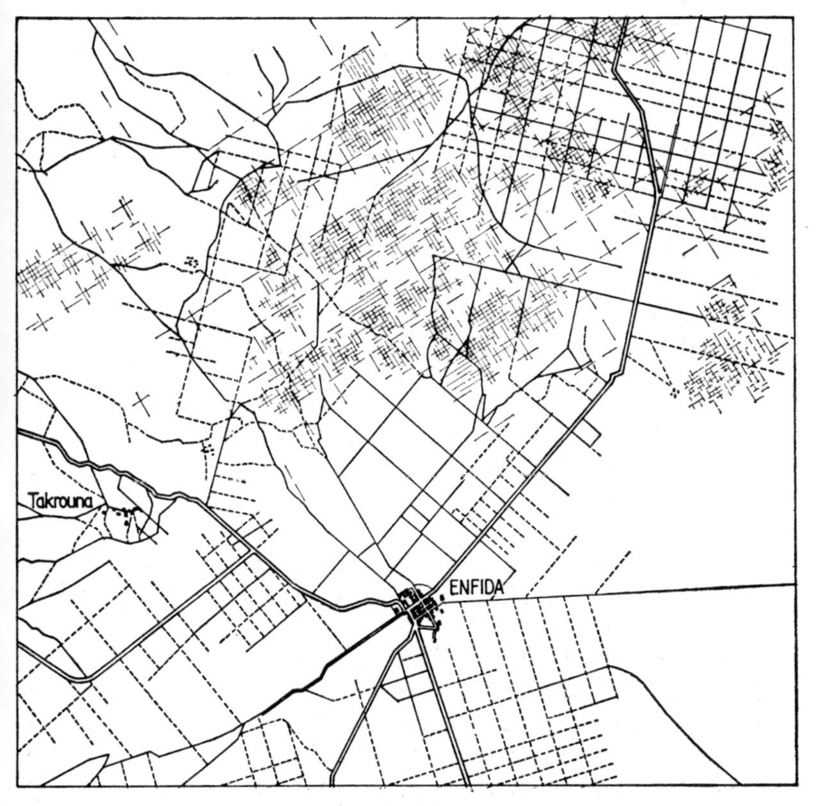

43. Centuriation and ancient land divisions near Enfida, Tunisia

The original province of Africa had its start in 146 BC, when the Romans razed Carthage to the ground. Some land division had already been effected by the time Gaius Gracchus founded New Carthage in 122 under the name of Junonia (p 182). The enormous allocation of 200 *iugera* for the first time made one holding, at least for certain privileged settlers, equal to one 'century', and the large number of settlers expected meant that a very large area had to be centuriated. Centuriation in the province is mentioned in the 'Lex Thoria' of 111 BC.[41] This required that appeals for loss of land consequent upon requisition should be settled within 150 days; and it excluded from centuriation the lands of Utica, Hadrumetum, Thapsus, Lepti Minus, Acholla, Uzalis, and Theudalis. Nevertheless, it is generally recognised that the Gracchan plan was not a success. Instead of *coloni* in the true sense, farming their holdings, the type of settler who went out tended to be more of a middleman who made money by buying and selling corn and other goods. From the point of view of surveying, there is little of interest for some 150 years after this foundation and the subsequent developments of the late second century BC, which resulted in about 15,000 sq km being divided into regular squares.

The scheme of AD 29 is known to us from nineteen centuriation stones found in southern Tunisia, sixteen of them near the Shott el Fedjedj, two a little to the north-west in the Bled Segui, and one to the north-east at Graïba.[42] These areas are in Africa Nova, made into a province in 46 BC. A number of them record that the centuriation was carried out by the third legion Augusta in the third proconsulship of Gaius Vibius Marsus, ie AD 29–30. The 'centuries' involved are all in the area DD VK, and the unusual feature is the extremely high numbering. The largest DD number is CXXXX, the largest VK number CCLXXX; in other words, there were at least 140 'centuries' to the right of the *decumanus* and at least 280 beyond the *kardo*. If they were of standard size, this puts the farthest centuriation stone found nearly 200km beyond the *kardo*.

We know something of the historical background to this operation. In AD 17, the Numidian Tacfarinas took to brigandage and harried the province for seven years until he was ambushed and killed. Tacitus[43] says that Tacfarinas sent envoys to Tiberius asking for lands, but that the emperor refused, as he put it, to subsidise brigandage. Evidently after Tacfarinas' death he decided that centuriation would solve some of the problems of that disordered area. We do not know to whom the lands were allocated, but it is quite likely that the whole scheme was only a skeleton one; at least, if it was not, the bulk of it has disappeared. As to the whereabouts of the KM and DM, we cannot be at all sure. Chevallier[44] rightly criticises the theory that the latter went through Ammaedara (Haïdra). At the time of this centuriation, Ammaedara was a military zone occupied by the third legion Augusta, which actually carried out the centuriation. It was only under Vespasian that it was founded as a colony, when it changed its function from military to civilian. In any case the main street of Ammaedara, although it went straight out through a central gate, was orientated differently from the adjacent centuriation as at present revealed. We should also beware of thinking that the scheme covered northern as well as southern Tunisia, as has been thought.

Heitland[45] wondered if the vast centuriation schemes of north and south Tunisia pointed to there having been an exodus of farming settlers from Italy on a massive scale. Fifty years of research since he wrote have tended to discount this possibility. Italian farmers were on the whole unwilling to go and settle in North Africa, in much the same way as French farmers, after France had acquired Algeria. Nevertheless, the lands continued to be allotted and taxed, and we are even able in certain parts to see with the help of air photography where individual trees, eg olives, were planted or where there were internal *limites*, as described on p 93.[46]

The emperor Hadrian took a personal interest in North Africa, bestowing the title of colony or *municipium* on a large number of towns. He pursued a policy of settling nomad tribes where possible, and legislated on uncultivated land. Under Septimius Severus we have details of an allocation south of the Shott el Hodna, Algeria.[47] Arable lands, pasture lands and springs were between AD 198 and 201 surveyed and allocated on the orders of Anicius Faustus, consular legate. The work was organised by the freedman Epagathus, obviously a Greek surveyor, and by the governor's *cornicularius* (centurion's adjutant) Manilius Caecilianus, and was carried out by a retired legionary of the third legion, Marcus Gennius Felix.

The centuriation in North Africa helped for hundreds of years with the fixing of tribute. Where the soil, through desert encroachment (a powerful factor in the fifth century) or for other reasons, had become unproductive, exemption from tribute could be granted. In AD 422 the total of lands thus exempted was reckoned as 5,700 'centuries' and 144½ *iugera* in Africa Proconsularis, and as 7,615 'centuries' and 3½ *iugera* in Africa Byzacena. A decree of AD 451, granting exemption from taxes to landowners expelled by the Vandals, mentions individual 'centuries' as still constituting the basis for taxation.

ASIA MINOR AND ADJACENT AREAS

There were at least twenty-five colonies in Asia Minor founded by Rome, but so far no centuriation has been reported. We have seen that the younger Pliny, as special commissioner in Bithynia, was keen for *mensores* to be sent out there from Italy. Some will have been required to measure and value buildings; others, assuming that there was no centuriation, were presumably needed to organise allocations and fix boundaries. From southern Turkey, boundary stones containing the name of the colony of Parlais have been found above the village of Bedre near Barla.[48]

From the province of Syria (Judaea), a centuriation stone is said to have been found near Haifa.

11

The Orange cadasters

We are fortunate in being able to study, partly in the original and partly in publication, substantial fragments of cadasters carved in stone at Orange (Arausio). The cadaster, in this sense, is a large-scale land survey carried out for taxation purposes. The derivation of the name, Fr *cadastre*, Ital *catasto*, is something of a mystery;[1] that given by most dictionaries, from medieval *capitastrum*, is rejected by two etymologists. Of the alternatives proposed, perhaps the most likely is from Greek *katastichon*, 'line by line', used as a noun to designate Byzantine ledgers.

We have seen (p 27) that in Ptolemaic Egypt the system of land taxation led to extensive compilation of local records. British and American land taxation is organised differently, so that modern parallels are chiefly with France, Germany and other European countries. Since the bulk of the stones was discovered after World War II, since many are now lost to us, and since they are at present unique, it has seemed worth while to devote some space to examining them.

The Gallic tribe occupying the area round Arausio was the Tricastini. On their territory, perhaps about 35 BC,[2] a Roman colony, *colonia Iulia firma Secundanorum*, was established. As the name shows, this was a settlement for legionary veterans, those of the second legion Gallica, replaced in that year by the second legion Augusta. Under the Flavian emperors, perhaps in Domitian's principate, the town name was changed to *colonia Flavia Tricastinorum*. This, contrary to the original phrase, attached the name of the local tribe to the settlement. The change of name, as will be seen, reflected a change of land distribution.

From 1856 onwards isolated fragments of surveys were discovered in and north of the magnificently preserved Roman theatre at Orange. Systematic study became possible when from 1949 to 1951 numerous fragments were excavated in the area south of the Rue de la République and Place de la République. They were carefully examined by the late Professor A. Piganiol and others and the results fully published by him.[3] Unfortunately in October 1962 there was a serious collapse in a wing of the Orange museum, as a result of which many fragments were virtually destroyed and some badly damaged.

First of all there is an inscription of AD 77 (p 189) on three long ornamental slabs, which explains how an edict from Vespasian caused the local authority to take action. The Latin is given in the caption to the plate; we may translate it: 'The emperor Vespasian, in the eighth year of his tribunician power (ie AD 77), so as to restore the state lands which the emperor Augustus had given to soldiers of the second legion Gallica, but which for some years had been occupied by private individuals, ordered a survey map to be set up, with a record on each "century" of the annual rental. This was carried out by . . . Ummidius Bassus, proconsul of the province of Gallia Narbonensis.' For a private individual to appropriate *ager publicus* had been an offence since at any rate the time of the Gracchi (in fact according to Livy[4] ever since the agrarian law of Spurius Cassius, 486 BC, though modern scholars disbelieve this).[5] Both in the last century of the Republic and under the Empire there were repeated efforts to restore *ager publicus* to the State wherever it had been expropriated, and Vespasian was particularly active on this score. He also, as has been seen, sold or taxed some *subseciva* to gain more money for the Treasury.

The emperor's edict appears to be linked with the first of the three cadasters into which the fragments were able to be sorted out. But this Cadaster A, like its successors, was erected by the local authority, not the State. The inscriptions were fixed to walls of what seems to have been a long, narrow building parallel to and about 100m from the *scaenae frons* of the theatre. This building must have been the local *tabularium*, a smaller counterpart of the central one in Rome. Of the three sets of inscriptions, B, which was the largest (originally 5m 90 high in all and at least 7m 56 wide), is also the best preserved. The 'centuries' of Arausio's allocations are represented by rectangles in A, by nearly square rectangles in B, and by squares in C. The main roads of the centuriation, the *decumanus maximus* and *kardo maximus*, are shown between relatively wide parallel lines, occupying about one-sixth of the width of a normally shaped 'century'. The scale for 'centuries' works out at about 1:6,000, which may have been a fairly typical scale for large areas of centuriation. But the DM and KM are clearly given an exaggerated width, whereas non-aligned roads are shown as narrower lines, and other roads of the centuriation tend to be ignored. Topographical features included within the 'centuries' are rivers and streams, islands, and these roads not aligned with the centuriation, ie earlier ones still functioning. But as the inscriptions are local authority records, not topographical maps, the chief emphasis is on land status, occupancy, area of holdings, and rental.

The following abbreviations are among those used:

(1) TRIC RED, *Tricastinis reddita*, 'restored to the Tricastini'. This occurs only in Cadaster B. We may deduce from the revised title of the colony that some time under the Flavian emperors the Tricastini earned the right to call the colony after their own tribal name. The Roman treatment of natives in an area where a colony was founded varied enormously; at one extreme is Colchester, where they were booted out by veterans and their lands confiscated, at the other extreme they might receive Roman citizenship. In the Orange area the Tricastini were at first dispossessed. When they did receive a good measure of land back, much of it seems to have been the worst land, as Cadaster B shows. The altered status of the colony and the necessity to show in the records these lands given back to the Gauls account for the scrapping of Cadaster A.

(2) EXTR, *ex tributario*, 'withdrawn from tribute-paying status'. Whereas the Tricastini paid tribute, the legionary veterans to whom lands were allotted did not. The amount of land originally given to the veterans at Orange is unknown, but from the areas of later holdings, a number of which are divisible by $33\frac{1}{3}$ *iugera*, Piganiol conjectured that this might have been the allocation.

(3) REL COL (or either of these abbreviations), *reliqua coloniae*, 'remaining in possession of (or remaining to) the colony'. Lands not allocated to the veterans belonged to the community; they were often pasture-lands and could be let. At Orange the rental was never in kind, always in cash, usually 4 *asses* ($\frac{1}{4}$ *denarius*), a comparatively low rental.

(4) R P, *rei publicae*, 'State lands'. This occurs only in Cadaster A. By the time of Cadaster B, these State lands had no doubt been amalga-

mated with those of the local authority. Although not, like (5), so labelled, they may have ranked as *subseciva* and been sold by Vespasian.

(5) SVBS, *subseciva*. Both types (p 94) are encountered at Orange. The fact that much of the land lay near the Rhône resulted in a greater proportion of *subseciva*.

Rents. Annual rents are always quoted in *denarii*, usually abbreviated as below, and *asses* (A, probably for *aera*, bronze coins). In origin there were ten *asses* to a *denarius*. But in the 140s BC the *as*, which had fallen in weight, was re-tariffed at sixteen to the *denarius*. At Orange the terms arising in the old decimal subdivision persisted, so that fractions of a *denarius* apart from *semis* ($\frac{1}{2}$) and *as* ($\frac{1}{16}$) were expressed in *libellae* ($\frac{1}{10}$ *denarius*), *singulae* ($\frac{1}{20}$), and *terruncii* ($\frac{1}{40}$). The fractions go in descending order except for the *as*, normally either nil or one; but if neither resulted in an exact fraction, sometimes $\frac{1}{2}$ or $1\frac{1}{2}$ *asses* were given. The abbreviation signs, in order, are:

X	*1 denarius*
S	*½ denarius (semis)*
—	*1 libella*
=	*2 libellae*
≡	*3 libellae*
£	*1 singula*
T	*1 terruncius*
AI	*1 as*

Thus 11 *asses* is S–TAI, ie $\frac{1}{2} + \frac{1}{10} + \frac{1}{40} + \frac{1}{16} = \frac{11}{16}$. This is to some extent reminiscent of the ancient Egyptian system under which only unit fractions were used;[6] but the 2 and 3 *libellae* go beyond this.

Measurements. For country areas, IVG = *iugerum*, S = $\frac{1}{2}$ *iugerum*, £ (*semuncia*) = $\frac{1}{24}$ *iugerum*. For town areas, measurements were given in square feet.

Names. The names commonly given are those of the principal lease-holders, sub-lets being ignored. Roman names predominate, but there are also Gallic and Greek names. There is a higher proportion of women than one might have expected.

Orientation. According to Piganiol the orientation of the three cadasters is as follows:

Cadaster	*Top of Diagram*	*Top of Survey*
A	north	east
B	west	west
C	north	west

Unfortunately so little remains of Cadaster A that it could, as suggested by Oliver, be inverted over 180°.[7] His theory is that the surveys were erected on three different walls, in such a way that viewers facing south, west and north respectively would see the plans the right way round. In this way the terms 'to the left' and 'to the right' of the *decumanus* would have had more reality for the viewer. If his suggestion is correct, however, Cadaster A is likely still to have remained in position even when it was obsolete. It is to be hoped that more material will be discovered which will remove any uncertainties about the orientation.

Cadaster A. This is very fragmentary, but Piganiol reconstructed it thus:

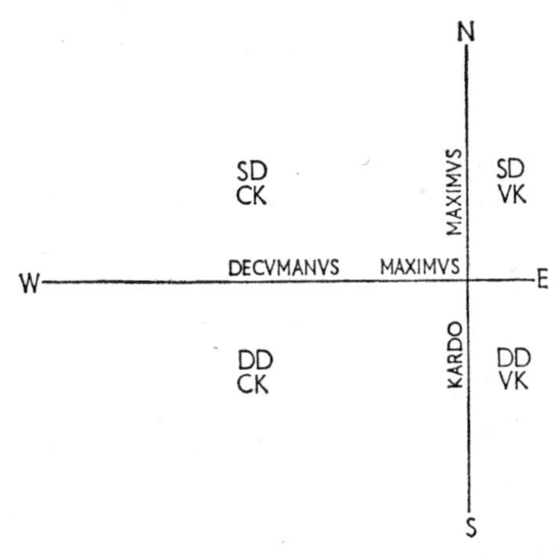

In the inscriptions of Cadaster A, the largest preserved total area of allocations in any one 'century' is 330 *iugera*, so that we must look for a size of 'century' as great as or greater than this. The 'centuries' were rectangular, so probably 20 *actus* north–south and 40 east–west, giving an area of 400 *iugera*. These are the same dimensions as in the example from Veturia near Emerita quoted by Hyginus (p 85). Fragment 7 (Fig 44) shows, near the intersection of the *decumanus maximus* and the *kardo maximus*, a braided river with what appear to be roads on each side, not straight but having a general direction at about 45° to the

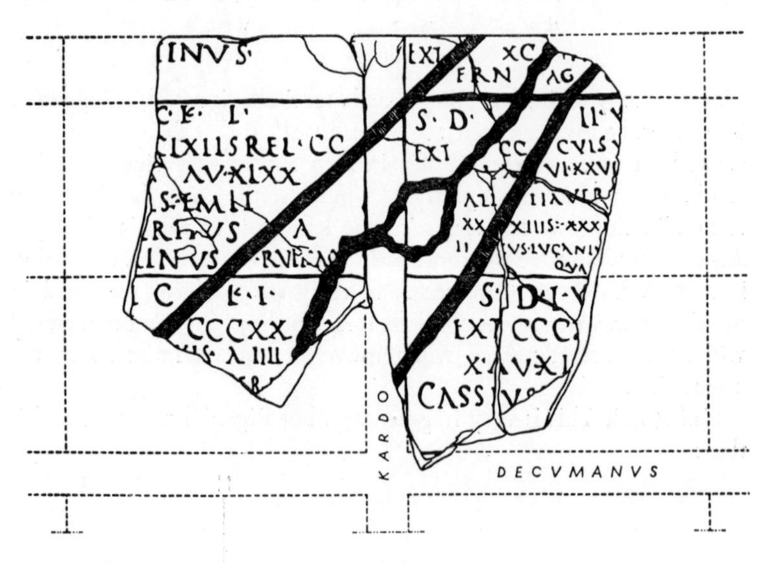

44. Orange cadaster A, fragment 7: area near intersection of KM and DM, showing a braided river with roads on each side

centuriation. Piganiol suggested that this intersection was fairly high up on the R. Aigue (Aygues), at the eastern edge of the Rhône valley north-east of Orange. Although that would explain why the area beyond the *kardo* (VK) appears to be restricted, it does not take account of the results of aerial photography. An aerial survey taken by M. Guy (Fig 45) shows a belt of rectangular centuriation extending from near Carpentras to just west of the Rhône WSW of Orange. Since this is the only rectangular area in the neighbourhood shown by air survey, all others being square, we must surely place the centre near the left bank of the R. Ouvèze at some point where it is flowing NE–SW; one possible location would be across the Ouvèze from Jonquières.

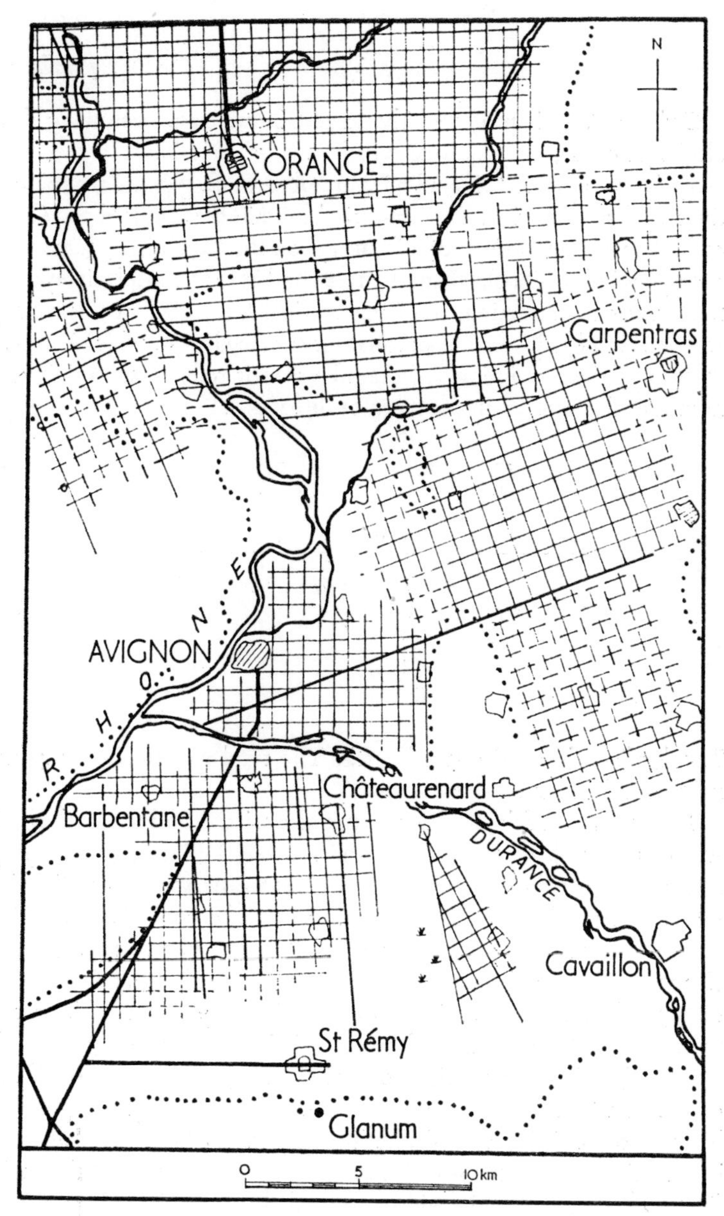

45. The Orange–Avignon area of the Rhône valley. Remains of centuriation. About eight different systems are seen, some superimposed. The rectangular system south of Orange, extending from near Carpentras to west of the Rhône, is likely to correspond to Cadaster A. Cadaster B is almost certainly the square system stretching north of Orange; only the southern part of this is visible on the present map

Cadaster B. This produced the largest inscription, occupying four blocks of stone, and the one which was when excavated the most extensively preserved, including a number of fragments mapping a river within 'centuries'. The 'centuries' contain the standard 200 *iugera* each. The average height of a 'century' on the stones is 14·3cm, their average width in Blocks II and III 11·85 and 12·04cm respectively; but there must be distortion, since the original 'centuries' in this survey were obviously square. The original survey faced west, which is also at the top of the inscription. The numbers on it refer to the minimum number of 'centuries' in the relevant directions.

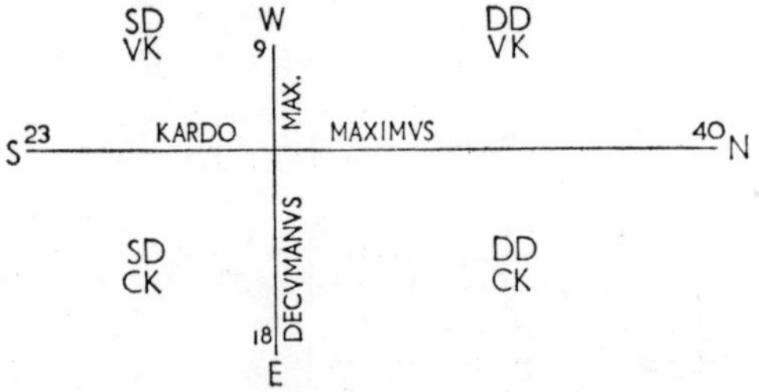

The *kardines* face $5\frac{1}{2}°$ east of north. The total extent is over 44km north–south, over 19km east–west.

The area west of the Rhône was not centuriated, and to east the land was in many places too mountainous. Tablet IVF (Fig 46) certainly seems to depict the Rhône, with *subseciva* to the west of it. Piganiol seems to have been right in thinking that the course of the river corresponds with modifications to the present course through the Gorge of Donzère, from about 500m north of the bridge at Touchelaze to beyond Viviers. This bridge (p 175) and its approach road from the west are exactly on the line of the centuriation as outlined by Piganiol. On the inscriptions the river occupies only about one-fifth of the height of a 'century', whereas its present width would correspond to between one-third and one-half of the height. We have seen that there is distortion, so that the scale is not to be relied on, and that the main *limites* are exaggerated in width. Nevertheless there is perhaps a presumption to be drawn from the tablets that in this area the Rhône has increased in width appreciably since the first century AD.

Piganiol, thinking it likely that the *decumanus maximus* is identical

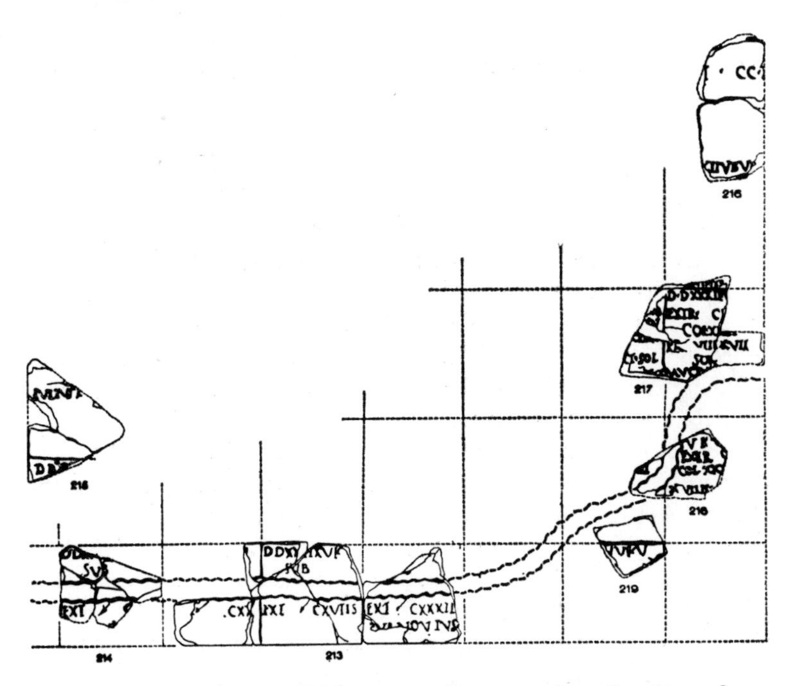

46. Orange Cadaster B, IV F, 213–219. Reconstruction of position of almost adjacent tablets which show the course of a river

with that of Cadaster A (but see above), placed the centre of Cadaster B south-east of Lapalud, about 700m south of Pont de Coucau, between Petit Galap and Grand Galap, some 19km NNW of Orange. This may well be right, but his Fig 36 should show the southern limit of Cadaster B farther south. It probably reached just south of Orange; for though we have no fragments corresponding to an area as far south as this, the air survey seems to point to it. The same scholar estimated that there were in Cadaster B about 1,700 'centuries'. This is surely an over-estimate, since there would only be 1,701 if the scheme formed a perfect rectangle, which because of the river and mountainous terrain it can hardly have. Inscriptions of this cadaster were found relating to 245 'centuries', the great bulk of which were in the northernmost two-thirds of the centuriated area. For the portions discovered, the percentages of allocation are:

	%
Assigned lands	53·1
Lands restored to Tricastini	36·2
Lands 'left over' for colony	10·7
	100·0

It will be seen that the amount of land retained by the colony under this survey was between one-ninth and one-tenth: Piganiol was inexact in saying 'la colonie garde 1/6'.

The lands restored to the Tricastini are all, as far as our knowledge goes, in the area DD CK, ie in the largest 'quarter' of the centuriation. Although its extent is large, it does not, as was pointed out by Piganiol, contain a great proportion of good land. Some, with the layout suggested, is on high ground between the middle stretches of the Berre and Lez valleys, some a mountainous area between the Berre and the Jabron. As to assigned lands, these are mainly on low, fertile land, as in the Rhône valley between the lower stretches of the Berre and the Lez; though assigned 'centuries' are also to be found even in the heart of the Tricastine territory. Communal lands are infrequent, and totally absent in Tricastine areas. It certainly looks as if the Roman descendants of the original settlers were favoured at the expense of most of the locals.

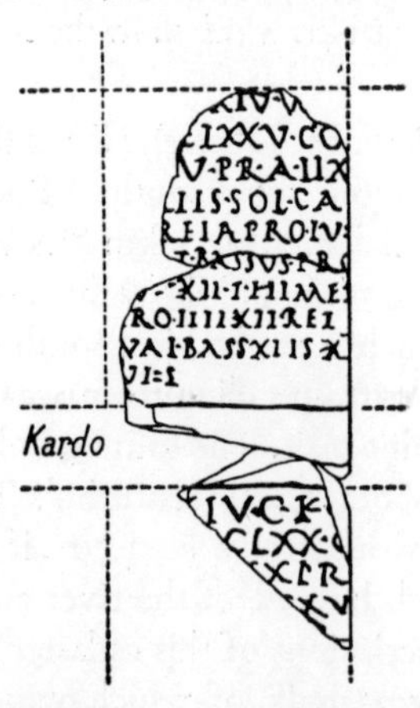

47. Orange Cadaster B, III C 171. Tablets representing sections of centuriation on each side of the *kardo maximus*: inscriptions and translations are in the text

As a well-preserved specimen of 'centuries', one on each side of the *kardo maximus*, we may take Block IIIC, No 171 (Fig 47). The expanded Latin is as follows:

sinistra decumani XIV ultra kardinem I: ex tributario CLXXV (iugera), coloniae XXV (iugera), praestant aera IIX, denariis XIIS; solvit Careia pro IV (iugeribus) et quadrante denarios II libellam et terruncium, Bassus pro IV (iugeribus) et quadrante denarios II libellam et terruncium, item et pro IIII (iugeribus) denarios II; reliqua Valerio Basso XII (iugera) et semis, denarios sex, duas libellas et singulam.

(KM)

sinistra decumani XIV citra kardinem I: ex tributario CLXX (iugera), coloniae XXX (iugera), praestant aera IIX, denariis XV; solvit. . . .

Translation:

Left of *decumanus* 14, beyond *kardo* 1.
Withdrawn from tribute-paying lands, 175 *iugera.*
In possession of colony, 25 *iugera.*
Tariff 8 *asses* per *iugerum*, total rent 12½ *denarii.*
Rents payable:
  Careia, for 4¼ *iugera*, 2 *denarii* 2 *asses.*
  Bassus, for 4¼ *iugera*, 2 *denarii* 2 *asses.*
  Bassus, for 4 *iugera*, 2 *denarii.*
Remainder to Valerius Bassus, 12½ *iugera*, 6 *denarii* 4 *asses.*

Left of *decumanus* 14, this side of *kardo* 1.
Withdrawn from tribute-paying lands, 170 *iugera.*
In possession of colony, 30 *iugera.*
Tariff 8 asses per *iugerum*, total rent 15 *denarii.*
Rents payable. . . .

The first of these two 'centuries' is thought to be at Nouvel Ilot, the north-east part of Ile St Georges; its north-east corner is today on the bank of the Rhône, 5½km north of Pont St Esprit. In Roman times it must have been more fertile than today, apart from a strip on the east, occupied by the river.

It should be added that the width of the *kardo maximus* as given on the stone is equivalent to about a quarter of the side of a 'century'. If the inscription were truly to scale, this would indicate a *kardo* width of 175m, which is impossibly large. Conversely, between other 'centuries' not separated by a KM or DM, there is only a single line, whereas in

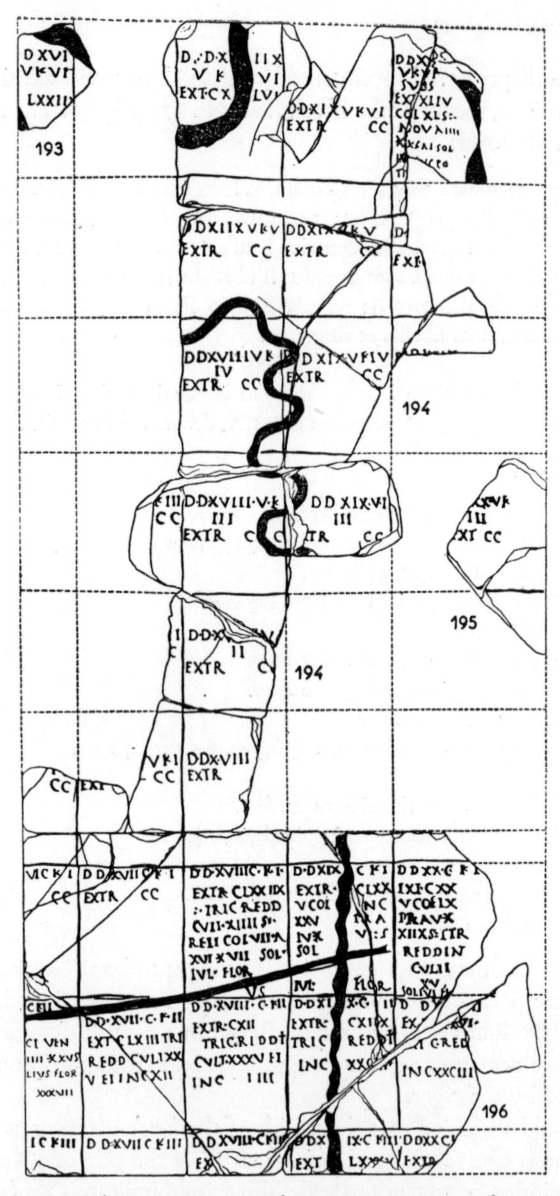

48. Orange Cadaster B, 193–196. This section contains a tributary of the Rhône, crossed in tablet 196 by an old road evidently curtailed by the centuriation. The latter extends from CK III to VK VI, including the KM, and from DD XVI to DD XX

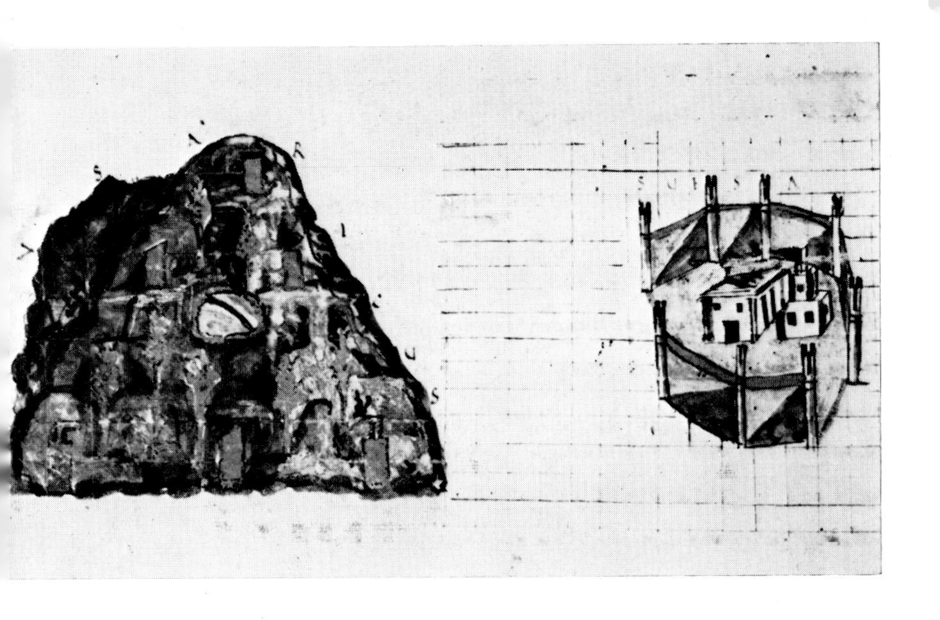

ge 171 (*above*) Mount Massicus (ARICVS incorrectly) and Suessa Aurunca, Campania, with ıturiation. The text makes it clear that at least some of the allocation was in strips, so that the gram cannot be entirely correct. Miniature (Agennius Urbicus) from MS A, Wolfenbüttel; *low*) Augusta Taurinorum (Turin) centre, and Hasta (Asti), right, with other places less easily ntifiable; roads, rivers, mountains. Miniature (Hyginus Gromaticus) from MS P, Vatican, intended illustrate entries which the surveyor will make on his maps. The corresponding diagram in MS A is far simpler

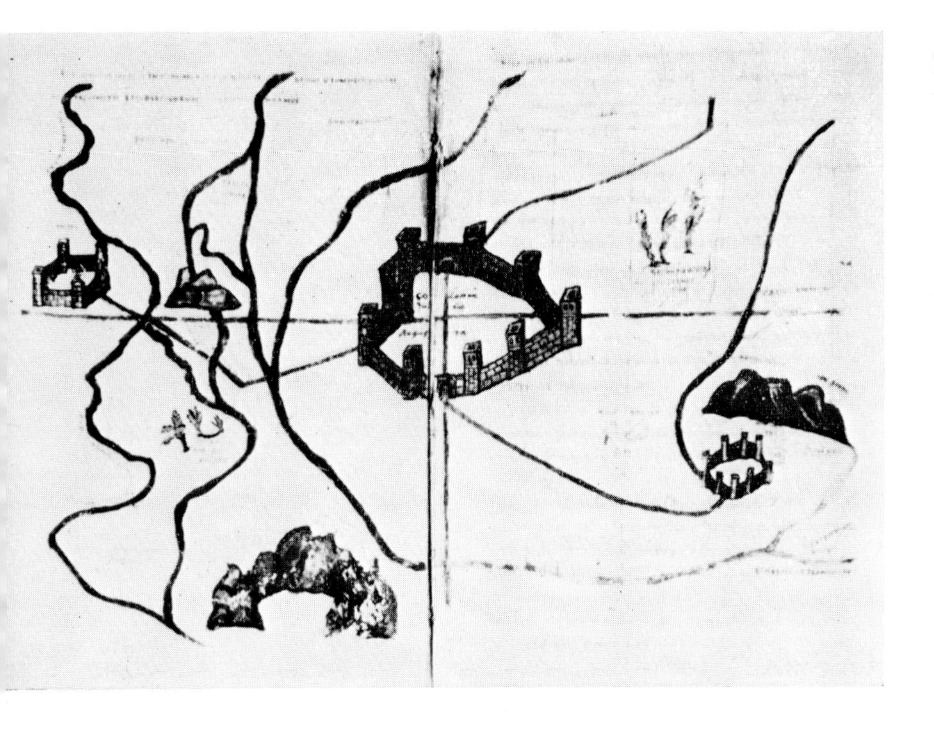

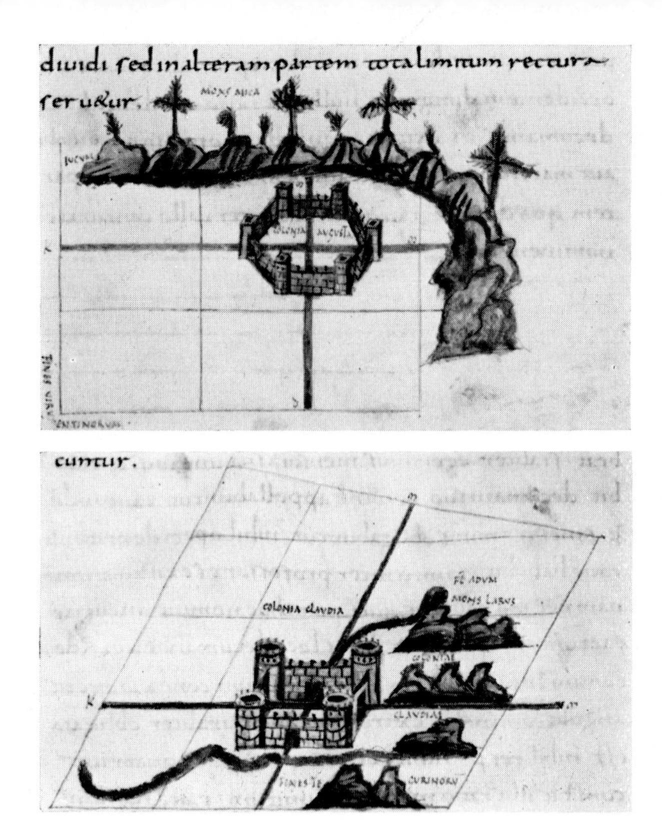

Page 172 (*above*) Two miniatures from the same page (Hyginus Gromaticus) from MS P, Vatica
(*top*) Augusta Praetoria (Aosta): centuriation cut off by mountains in two directions. No centuriati
in the Aosta valley has as yet been discovered; (*middle*) Colonia Claudia, presumably Aventicu
(Avenches, Switzerland), a colony founded by Vespasian; CLAVDIA could be a corruption
FLAVIA. River, mountains, centuriation, boundary of the Tegurini; (*below*) common pasture la
between two centuriated areas. Miniature (Frontinus) from MS A, Wolfenbüttel

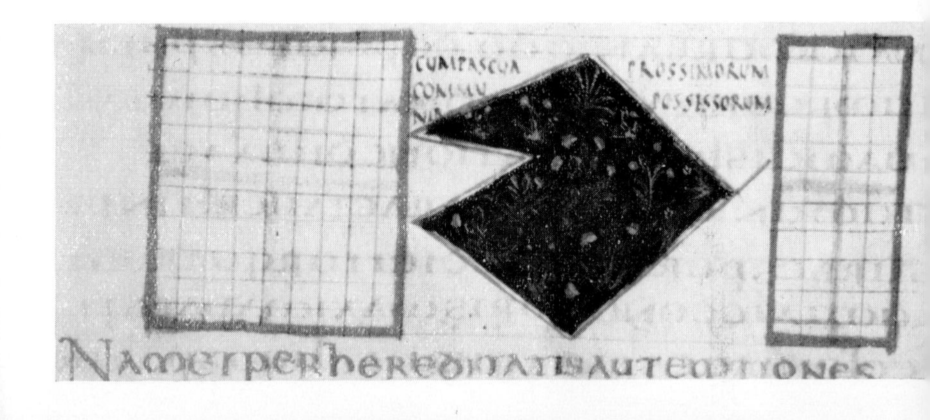

reality each 'century' must have been separated from the next, as elsewhere, by a strip of land (*limes*) of between 8 and 20 Roman feet.

Cadaster C. Here the original survey faced west, but the inscription has north at the top.

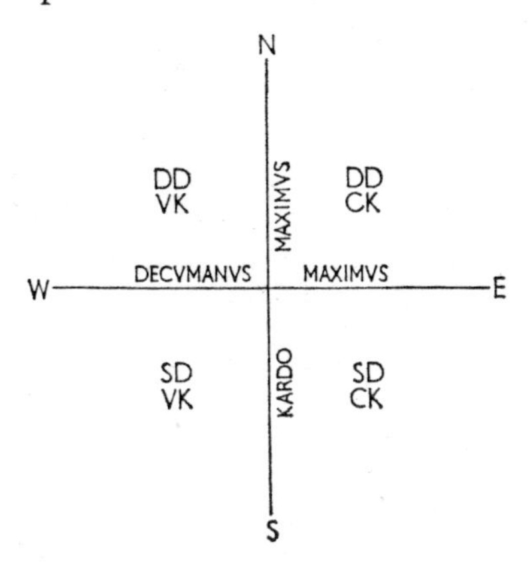

Although fragmentary, it preserves parts of two areas: (1) a central area, (2) a western portion containing islands in the Rhône, the *insulae Furianae* (Fig 49). These extended about 5km north–south and over 2km east–west. It is evident that both form part of the same survey; thus, each has the DM 3·5cm wide. The total number of 'centuries' east–west is about 38, north–south about 21. The average size of a 'century' on the inscription is 12–12·5cm square, a scale of approximately 1:6,000. Unfortunately the course of the Rhône and the positions of the islands in it have so changed over 1,900 years that attempts to locate the survey exactly are bound to be speculative. Piganiol suggested that the DM might have been near the Rhône about 1km north of Roquemaure, with the KM some 6km west of Carpentras. The latter he orientated about 7½° west of north; but the air survey seems to show the orientation to be more like 10° west of north, ie even more at variance with that of Cadasters A and B. One inscription[8] records a 'century' as owned by Quintus Curtius Rufus (the same name as the historian), who is called the *inventor* (founder) of the place.

In addition to fragments from Cadasters A–C, fragments from the public record office of Orange, the *tabularium publicum*, were found. We know that such offices were not uncommon; in neighbouring areas

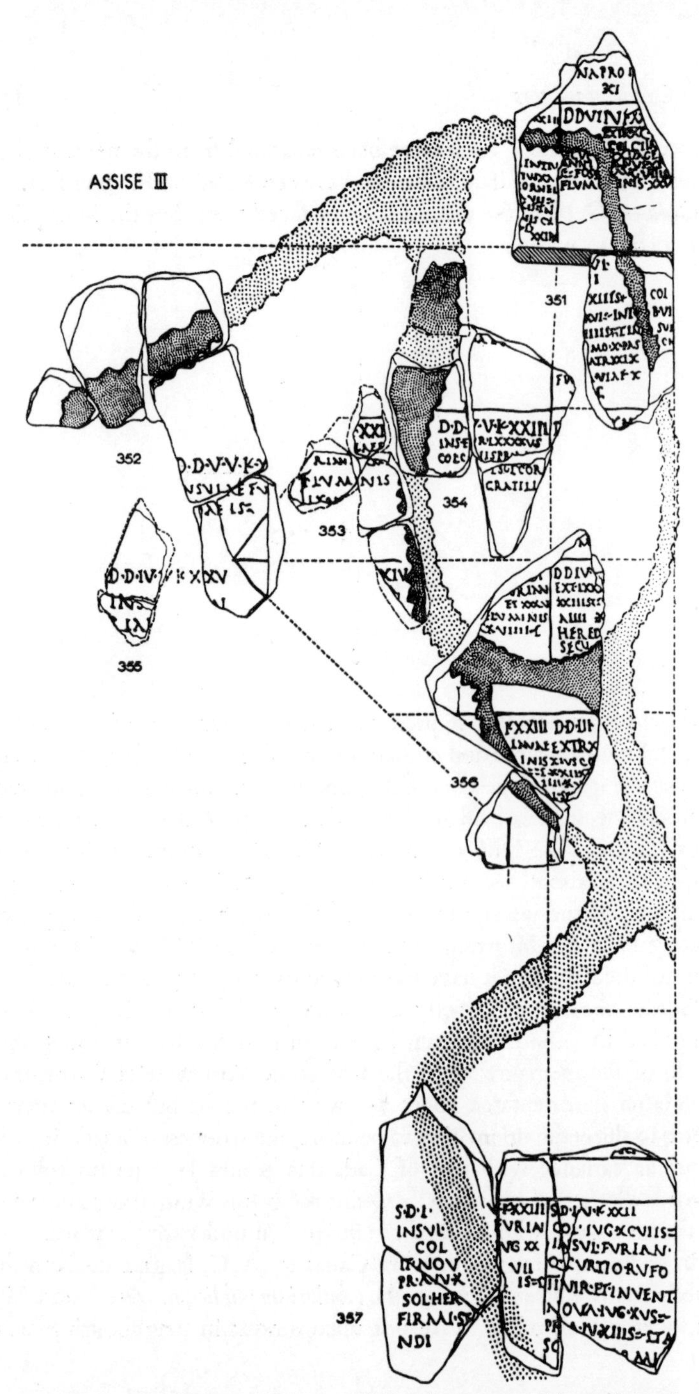

49. Orange Cadaster C, 351–357. The *insulae Furianae*, islands formed by the Rhône, which were centuriated in the ordinary way

they are recorded at Vienne and Vaison. The chief concern was to record arrears due to the community, on which interest was charged. These were entered in the public ledger of debts, the *kalendarium*, named after the kalends, when debts were due. A number of them relate to fines for encroachment on public domains or on *areae*, public or sacred areas such as the city wall surround and temple surrounds. Presumably surveyors would be concerned with these whenever a dispute arose how far the protected area extended. Some inscriptions of the record office mention the letting of stands, called by a Greek word *merides*, literally 'lots'. Permanent stands were let out at 1 denarius a foot.

Items now visible in the public rooms of the Orange museum include: (1) the inscription recording Vespasian's edict, (2) some fragments from Cadaster A, including the central area, (3) on the north wall, a rectangle of Cadaster B 25⅓ 'centuries' high and 39 'centuries' wide, including the central area, which is to be seen half-way up the wall and seven 'centuries' from the left, (4) two fragments from Cadaster C, (5) a number of fragments from the public record office.

To summarise, we do not know anything about the original centuriation of about 35 BC, except for the likely size of the allocations. If it was the same as Cadaster A, it was not on the same orientation as the colony of Orange, whose *kardo*, going through the famous triumphal arch, is 15½° west of north, different, surprisingly enough, from all the known centuriation schemes. Cadaster A, the only one in which State lands are listed, is clearly the oldest and is likely to be connected with Vespasian's edict, even if based on an earlier layout. The State lands may have been presented to the colony by Domitian, who is known to have done this elsewhere. Cadaster B may have been drawn up under Trajan. It shows the greatest use of allotted land, developing under the changed status of the colony as a Gallic town, even if there was some gerrymandering of the allocations which resulted in the Tricastini getting poorer lots. It is tempting to think that a bridge over the Rhône may have been built in Trajan's principate. The present suspension bridge at Touchelaze, which has wooden planks and is confined to one-way traffic, and its approach road from the west, occupying between them 1·1km, are exactly on the probable line of Cadaster B. East of the gorge the centuriation must have been broken off at the cliff line. Cadasters A and B follow roughly the direction of the Rhône valley from Montélimar to Mondragon and beyond, while Cadaster C reflects the curving of the Rhône south of Orange and the widening out of the plain. Air survey has established the existence of other centuriation patterns south of Cadasters A and C, extending beyond Avignon and across the Rhône south-west of Orange. No inscriptions seem yet to have been found which relate to these patterns.

When we turn to the present-day map, we find that the lines of internal *limites* within the 'centuries' have been preserved more often than the external *limites* (p 190). This could be due to the persistence of features like the cypress hedges, regularly planted in the east–west lines as a protection against the north wind (Mistral), which are such a prominent feature of the area. Today it is a region of arable cultivation, especially of vines, fruit-trees, and maize, on the lower ground, with pasture on the upland. In general the searcher after traces of centuriation will be disappointed at the scanty remains, and it is chiefly the inscriptions that enable us to make, with the aid of air photography, a hypothetical reconstruction.

The quality of execution demands some comment, since in the absence of surveyors' maps of the period it is of interest to see how carefully or otherwise the work on this nearest parallel to them was carried out. On the question of cartographic accuracy, whereas both Cadasters B and C should have been carved with exact squares, the latter, as has been seen, is rectangular. Moreover the width of the main intersecting roads is exaggerated, whereas other *limites* are depicted simply by a straight line, without allowance for width. As to the rivers, Piganiol maintained that Cadaster B, allowing for the distortion in its rectangles, represented a large part of the course of the R. Berre very accurately. This is less likely than that of the Rhône to have changed over the centuries since Roman times. The writing is carefully executed, but to some extent lacks uniformity. The letter A, for instance, appeared in the various inscriptions as follows:

Cadaster A
Fragment 47
Subseciva
Cadaster B
Cadaster C
Islands

No chronological order stands out from a comparison of the scripts. The lines are very regular; and in the case of the Vespasian inscription (p 189) the entire execution, complete with egg-and-tongue and other mouldings, is excellent. Even if Augustus' principate is the period of greatest epigraphic perfection, that of Vespasian is little behind. When we come to the squares and rectangles representing centuriation, there is often so much to squeeze in, even with drastic abbreviations, that regularity has sometimes to be sacrificed to practical considerations. But all in all we can only admire the craftsmanship.

12

Colonies and State domains

Ancient colonies were not like those of relatively modern times or their successors. Phoenician and Greek colonies were mostly maritime settlements designed to promote trading, the Greek *apoikia* meaning 'place away from home'. The Latin word *colonia* means a settlement of *coloni*, derived from the verb *colere*, 'to till', hence literally 'tillers of the soil'. But as Salmon[1] puts it: 'When used technically the word had a precise significance. It denoted a group of settlers established by the Roman state, collectively and with formal ceremony, in a specified locality to form a self-administering civic community.' We shall see that much of the earlier Roman colonisation had a military or naval defence function. A letter from Philip V of Macedon to Larisa[2] attributes Rome's power to her system of colonisation and her liberal citizenship policy. Nevertheless, the basic agricultural background was always kept in view and land surveyed accordingly.

The land used for establishing colonies was mostly *ager publicus*, state domains annexed from defeated enemies. As a result, the Corpus tends to equate centuriation with the assignment of *ager publicus* given to colonies, though in fact this is an over-simplification. A colony could exceptionally be set up on land that was not state domain, and the territory centuriated could be that of a *municipium*, a town incorporated into the Roman state with or without Roman citizenship.

Colonies began to be founded at an early stage in Roman history. Three, Fidenae (Castel Giubileo), Cora (Cori), and Signia (Segni), are said to have been founded under the kings, obviously to protect Rome against hostile tribes threatening her in the Tiber valley or on the hills to the south-east. Eleven more of these 'early Latin colonies' had been founded before the dismemberment of the Latin League in 338 BC. They tended to be planned on hill-top sites which had some strategic importance. Sometimes they were founded in pairs which could help each other; thus Sutrium (Sutri) and Nepet (Nepi) were both started about 382 BC. This was not long after the sack of Rome by the Gauls had shown weaknesses in the defences on the North. We can only be sure of one maritime colony having been founded as early as this, Circeii (Monte Circeo), whose mountainous promontory provides a good look-out post over land and sea.

In 338, after a war lasting three years, Rome defeated the other members of the Latin league. The larger members became subject allies, the smaller ones were incorporated. Seven continued to be known as Latin colonies; they were privileged in that they did not have to give up any land and that any of their citizens who happened to be in Rome could vote there. As a protection against sea attacks, a pair of colonies of Roman citizens was founded at Ostia (Ostia Antica) and Antium (Anzio). Ostia, at the mouth of the Tiber about 25km from Rome, had been inhabited earlier, but from now on became Rome's main port. Excavation has revealed a fort as the original settlement, with an area of over 2ha (over 5 acres). It may have had 300 settlers, each with a plot of 2 *iugera* (0·504ha/1·246 acres), outside the walls of the fort and with pasture rights. The small number of settlers is explicable from the unwillingness of Roman citizens at this time to settle away from Rome. But one can show from ancient writers that the numbers 3, 30, 300 had something of magical significance. There were 30 communities in the old Latin League; the legend of the sow with a litter of 30 symbolised these; and 3, 30, 300 are significant numbers in Virgil.[3] The probable smallness of plots, well below subsistence level, has no doubt rightly been explained on political grounds: if the settlers were given bigger plots, they might have moved to a higher censorial class, with an impact on voting in the assembly. To earn their living, the settlers would either have had to undertake harbour or other duties or make extensive use of the common pastures which were normally available on the territory of a colony.[4]

The first colony for which we have a precise figure, 300 settlers with 2 *iugera* each, is Anxur-Tarracina (Terracina), likewise a colony of Roman citizens. This total occupied 600 *iugera*, which at the standard equivalent would occupy three centuries. Rome had defeated the

Volsci and was wanting to protect her communications at the point where the road reached the sea in a mountainous area. Although the colony at Tarracina was founded in 329 and the Via Appia, Rome's main artery to the South, was not built until 312, there must have been an earlier road on the same line in this area, since Hyginus Gromaticus tells us that at Anxur the *decumanus maximus* may be seen on the line of the Via Appia. The accompanying illustration in the Corpus (p 120) is discussed on p 116.

It is significant that from the foundation of Ostia and Antium down to 184 BC all colonies of Roman citizens, ten in all, were founded on a coast, at first that of the Tyrrhenian Sea, later on the Adriatic. The intention was to protect Rome from attack by sea, by means of patrols of *coloni maritimi*. During this period, however, Latin colonies were also being founded, some on the seashore, some inland. The numbers of their settlers were much higher, and these sometimes included Roman citizens, who thereupon lost their citizenship. Thus Cales (Calvi) was founded in 334 BC with 2,500 settlers, some 150km south-east of Rome. This was obviously a new departure, paving the way for Roman expansion on a large scale. The numbers increased: at Interamna on the Liris in 312 BC there were 4,000, at Alba Fucens (303 BC) 6,000. In 273 BC two Latin colonies, Cosa north of Rome and Paestum south of Naples, were founded to guard the coast. Salmon[5] makes a special study of Cosa and its territory. It seems likely that further excavation will confirm the view that there was a constant fight against the silting up of its harbour. During the First Punic War (264–241 BC) a number of colonies were founded both on the coast and inland. Shortly before the Second Punic War broke out (218 BC), a new pair of colonies was planned to control Cisalpine Gaul: 6,000 settlers were sent to Cremona, 6,000 to Placentia (Piacenza).[6] According to Livy there were by the outbreak of the war 30 Latin colonies and 10 colonies of Roman citizens. A majority of the colonies stood firm against Hannibal, and their loyalty played an important part in his defeat.

The period after the Second Punic War[7] witnessed an expansion north and south in the Peninsula. In order to attract settlers to the South on lands taken from the Bruttii, allotments of 15 and 20 *iugera* were made, yet even so not all the holdings were taken up. In Cisalpine Gaul Bononia (Bologna) was founded in 189 BC with allotments of 50 *iugera*. By way of inducement to settle, this and a number of other settlements were given attractive sounding names, Bononia from *bonus*, 'good', Placentia from *placens*, 'pleasing', etc. The following table shows how the number of *iugera* allotted to each settler varied in the colonies of Roman citizens founded between 184 and 181 BC:

Date	*Colony*	*Number of iugera*
184	Potentia	6
184	Pisaurum	6
183	Saturnia	10
183	Parma	8
183	Mutina	5
181	Graviscae	5

Allotments in Latin colonies vary much more.

An important factor in the planning of Cisalpine Gaul was the building of the Via Aemilia, under Marcus Aemilius Lepidus, consul 187 BC, whose name is perpetuated in the modern Emilia. The Via Aemilia led from Ariminum (Rimini) to Placentia, and much of the centuriated land adjoining it took its orientation from the road. In the north-east a Latin colony was founded at Aquileia in 183. The Senate decided that it could not be given the status of colony of Roman citizens because of its distance from Rome; as a result it was not until a road linking with the Via Aemilia had been built that enough settlers could be found. New settlers were sent there in 169 BC. In 173 BC (p 147) Romans were given 10 *iugera* a head, allies 3 a head, of *ager publicus* in Cisalpine Gaul. Between 173 and 124 BC perhaps only one colony was founded.[8] These were peaceful times in Italy, and the oligarchy chose to run state domains as large and profitable farms rather than distribute them in smallholdings.

Vast estates, *latifundia*, had grown up in many areas of Italy, run with very large numbers of slaves. The conditions of work on those in Etruria were so bad that they are alleged to have made Tiberius Gracchus determined to put an end to the system. The landlords of these large estates, if they were on *ager publicus*, ranked legally as *possessores*, which did not imply as great a degree of protection as full ownership. There was, however, protection against disturbance by a third party, provided the *possessio* was considered just. Relying on legislation doubtfully ascribed to the fourth century BC, Tiberius Gracchus, tribune in 133 BC, passed an agrarian law limiting the extent of state domains which could be held by a *possessor* to 500 *iugera*, with

250 extra for each of two children. On lands thus made vacant he proposed to settle smallholders paying rent to the State. Despite opposition, the three-man land commission consisting of his father-in-law Appius Claudius, himself and his brother Gaius, was able to operate; they did not cease operating when Tiberius was killed, but there was an intermission from 129 to 123/2. When Gaius became tribune in 123, he instituted a very large programme of colonisation. Colonies of Roman citizens were named after gods, Neptunia at Tarentum (Taranto), Minervia at Scolacium (Squillace) and Junonia at Carthage, proposed by another tribune acting for Gaius. Opposition to this last colony, planned to take 6,000 settlers from Italy with 200 *iugera* apiece, was intense. The soil of Carthage, it was argued, had been cursed by Scipio; Romans who settled there would be cut off from Rome; a new powerful Carthage might arise in opposition to Rome. While Gaius was visiting the site, opponents circulated stories that the gods had opposed colonisation on cursed land, causing a storm, that the omens were bad, and that boundary stones had been taken by wolves, which would be surprising as wolves did not exist in North Africa.[9] The whole Junonia scheme, with its vast, flat extent and its pioneering spirit, has been compared by Chevallier[10] with the settlement of the American Far West. Clearly Gaius Gracchus did not see the scheme through, since he stayed only ten weeks; and in the end the foundation of the colony was cancelled, though it seems that individuals were allowed to retain their holdings, which were centuriated either then or shortly after. There was certainly land enough to provide 200 *iugera*, a whole century, for each immigrant; but that may have been the maximum, allotted only to *equites*, otherwise, including the existing communities, the whole scheme would have covered over 2,000 sq km. But Gaius Gracchus had also to contend with outbidding by a political opponent. In 122 Marcus Livius Drusus proposed 12 new colonies, each of 3,000 settlers taken from the poorest classes in Rome; how many, if any, were founded is uncertain. Gracchus' death in a riot put a temporary halt to the more grandiose plans.

Colonies outside the Italian peninsula, however, were destined to multiply. An early example is Narbo Martius (Narbonne), founded about 118–14 BC despite senatorial opposition and named like the Gracchan colonies from a god, Mars. Refounded by Julius Caesar for the veterans of his tenth legion, it became the capital of Gallia Narbonensis. The Emperor Claudius also refounded it, and by the process of agglomeration its name by then became Colonia Iulia paterna Claudia Narbo Martius decumanorum. Some land outside Italy was centuriated and mapped without a colony being founded: Corinth, sacked by Rome, as was Carthage, in 146 BC, was ordered in the agrarian law of 111 BC[11] to have all its land measured, and boundary stones erected. The same law[12] defined the status of land in Italy and North Africa, declaring holdings of *possessores* within the 500–1,000 *iugera* limit to be private land, and provided that claims for loss as a result of assignation of land should be settled within 150 days and noted on the surveyors' maps.

In the first century BC the main purpose of founding colonies was to provide land for legionary veterans. Land for those in the Italian peninsula was either purchased or confiscated, and more colonies were founded elsewhere. Sulla, relying on his powers as dictator, ordered that land taken from the towns that had opposed him in the civil war of 83–2 should be centuriated and gave it to his veterans. Julius Caesar provided land in Campania in 59 BC for Pompey's veterans from the East. From Cicero's letters one can see how worried people in the countryside of Latium were at the possible threat of confiscation. When Caesar became dictator after the civil war of 49–48, he founded a very large number of colonies, mostly outside Italy: from the figure of 80,000 *coloni* mentioned by Suetonius as settled in the provinces, one may guess a figure of about 30–40 new or re-founded colonies there. Urso (Osuna, Spain) was officially called Colonia Iulia Genetiva Urbanorum. Bronze tablets containing part of its charter are preserved in the National Museum, Madrid. The settlers came from Rome and included freedmen, who were eligible, according to the charter, for election to the local senate. A small tax was levied on holdings in this and other colonies in the provinces. The two rivals of the past, Carthage and Corinth, were made into colonies by Julius Caesar.

The most extensive founder of colonies, however, was Octavian (Augustus). The following figures are available:

Soldiers demobilised after the battle of Actium, 31 BC (many settled in colonies)	over 120,000
Colonies founded or re-founded in Italy	28
Colonies founded or re-founded in the provinces (there may have been many more)	over 80

Paid in compensation for land in Italy	600 million sesterces
Paid in compensation for land in the provinces	260 million sesterces
Total of provinces in which colonies were founded	9 (plus Mauretania)

Most of these figures come from Augustus' own *Res Gestae*, which like most propaganda autobiography omits what is unpalatable. Thus after Philippi, 42 BC, the Second Triumvirate (Antony, Octavian and Lepidus) confiscated some lands without recompense. The disgust felt by the peasant in Virgil's *Eclogues* at being evicted from his ancestral lands near Mantua to make way for an uncouth soldier is very real. The legionaries themselves complained, as we learn from Tacitus:[13] 'Anyone surviving the hardships of legionary service was dragged off to all sorts of countries, where under the name of "lands" they were given swamps or barren mountain tracts.' Even the system of drawing lots clearly did not satisfy the grumblers. Augustus exempted a number of colonies in the provinces from taxation of holdings; but at Arausio (Orange) we find Vespasian over a century later seeing to it that the taxes do not lapse. Augustus' colonies, like some of the early ones, often had a military purpose: thus Augusta Praetoria (Aosta) was founded in 24 BC with 3,000 veterans of the praetorian guard, on the camp set up by Varro the previous year for his victorious campaign against the Salassi.

It is clear from the Corpus that the usual size of holding in the early Empire was 50 or 66⅔ *iugera* (one-quarter or one-third of a century, 12·6 or 16·8ha). The Emperor Claudius, himself born at Lyon, was keen both on extending Roman citizenship and on founding colonies in the newer provinces. Cologne, which has retained the first word of his title Colonia Claudia Augusta Ara Agrippinensium, was founded by him in AD 50. In Britain, whose conquest he had himself supervised, he founded in the same year the colony of Camulodunum (Colchester, Colonia Victricensis), planned as the capital of the new province, the site being dominated by an impressive temple. Like Augustus, Claudius founded a number of colonies in Mauretania. Vespasian (AD 69–79) raised several places in Gaul and elsewhere to colonial status, Trajan (AD 98–117) founded colonies in North Africa and on the Danube, and Hadrian (AD 117–38) in various areas.

After Hadrian it became common, as had been done to some extent under earlier emperors, to confer the title *colonia* on certain existing towns in the provinces. This made it easier for their leading men to become members of the Roman Senate, and also gave some hope of relief from taxation.

The activity and presence of surveyors whenever a colony was founded has already been stressed—it was clearly a most important part of their work. Yet the growth and decline of Roman surveying does not correspond to the growth and decline of colonisation. We have seen above (p 44) that there is evidence of much surveying activity after the principate of Hadrian, whereas new colonisation was rare. The explanation of this is that in the later Empire there was a constant growth of bureaucracy, and that surveyors like other civil servants found themselves involved in paper-work, legal disputes, etc, on a large scale. The Corpus is full of references to colonies and their territory, and it is clear that surveyors under the Empire, even if not always under the Republic, were responsible for many of the routine matters connected with the territories of colonies. The successive Orange surveys must have required many man-hours, and Trajan's colonisation was presumably in the forefront of his schemes, though it was chiefly his building programme in and near Rome that kept his surveyors too busy to be sent to Asia Minor, where the younger Pliny was hoping for them.

The brief account of colonies given in this chapter has mainly been given in chronological order. In the Corpus we find such a description only in very limited form. Hyginus Gromaticus gives a brief historical account concentrated on the periods of Julius Caesar and Augustus. Elsewhere he gives examples of features from colonies as early as Tarracina (*decumanus maximus* equated with Via Appia) and as late as Admedera (crossroads in centre of colony). But we have also in the Corpus systematic lists of colonies with certain facts about them that would have been useful to surveyors.

The *libri coloniarum* are in the form of catalogues arranged geographically in two lists: (1) separate areas of Italy, including Sicily, followed by a fragment on Dalmatia, (2) separate areas of Italy only. One imagines that similar lists existed in antiquity for other parts of the Roman world also. The collection comes from various sources, as is made clear by extracts labelled 'from the book of Balbus', 'from the book of regions', 'from the commentary of the Emperor Claudius',[14] etc. The manuscripts of the first *Liber coloniarum* have different arrangements, but Mommsen[15] considered the original subdivision to have been:

1 Picenum
2 Valeria, the late Empire name of the area of central Italy formerly constituting the fourth region (p 99)
3 Tuscia (= Etruria)
4 Umbria
5 Campania (including Latium)
6 Samnium
7 Lucania
8 Brittii (= Bruttii), roughly corresponding to the modern Calabria
9 Apulia
10 Calabria; in ancient times this was the heel of Italy
11 Sicily

This corresponds only approximately to the eleven districts of Italy (p 99), but exactly to the 'provinces' into which Italy was divided in the fifth century AD. Within the divisions of Italy, the order of colonies is mostly alphabetical.

The type of information tends to vary from district to district. We may be told the colony's status, its founder or founders, usually giving an indication of date (but there is no historical information earlier than the Gracchi), in some cases its full name and whether the settlers were military or civilian. Then we may be told something of the orientation of its land, the size of its 'centuries' and in a few cases the total area of the land. If there is non-centuriated land of unusual shape, this may be mentioned. There is quite a lot on boundary markings, whether of conventional stone (marble, limestone etc) or flint or wood, sometimes specified, eg olive-wood; in one case millstones act as boundary stones. Inscriptions on boundary stones are rarely mentioned, but we are often told the distance of these from each other, including any irregularities. There are frequent entries on the legal servitude on land each side of the public highway, as explained on p 93. Other legal stipulations may also be given. In a few cases there is a record of a river or stream constituting the boundary. Details of ownership are very rare—we learn, for example, that part of Lanuvium belonged to the Vestal Virgins.

The second *liber coloniarum* is very fragmentary except for Picenum, on which it has rather more information than the first, though with much overlap. Thus, in the first *liber coloniarum*, the entry on Asculum Picenum may be translated: 'The land of Asculum has been allocated in various parts with internal [subsidiary] *limites*. It was measured out with boundary marks of the emperor Claudius, made like little boxes, and with others, wooden ones at which sacrifices were made. The distance between these boundary marks is 1,200ft or less. But it remained unsurveyed, and the courses of neighbouring streams [?] are kept as boundaries. Its land was allocated to soldiers; but some areas were not included in the allocation.' The second *liber coloniarum* adds a note on the types of boundary stone, (a) red sandstone, (b) cairns, and on two land measurements in the area, in each case with a military connection. Asculum Picenum (Ascoli Piceno) lay on the R. Truentus near the Adriatic but among high mountains. Punished by the Romans for its rebellion in the Social War, Asculum was colonised by the Triumvirs or Augustus.

The greatest gap in the *libri coloniarum* is any knowledge of what occurred before 133 BC. A compilation of the fifth century AD, they incorporate some records of the early Empire. If we had no other records on colonies, we should not realise the vast amount of colonisation carried out earlier. Thus we read that Liternum in Campania was a colony founded by Augustus; this may be true (it could have been one of the 28 which he claimed to have founded in Italy), but we do know from other sources that it was founded as a colony in 194 BC. Mommsen was able to point out many mistakes in the catalogue. It has double entries, Asetium and Casentium (both unknown), Anagnia (genuine) and Calagna (spurious),[16] and numerous other corruptions. It calls Alatrium a colony and Abella a *municipium*, whereas Alatrium was never colonised and Abella was. There are many other errors, and they induced Mommsen to place a very low value on the work. Later researches by Pais[17] and others tended somewhat to rehabilitate it. Quite often there is more omission than error; thus, some colonies were repeatedly refounded, whereas the *libri coloniarum* record only one foundation.

The verdict must be that they preserve both much truth and much falsehood, and that as a source for the activities of the land surveyors they must be used with caution.

13

Accurate planning in Roman Britain

Attempts have recently been made by Alexander Thom and others to show that the megalithic monuments of pre-historic Britain bear remarkable witness of precise measurement and astronomical observation. In this chapter, however, only the period from the Roman conquest onwards will be discussed.

CENTURIATION AND PARCELLING

When it was realised what vast areas of Italy and Tunisia were centuriated, it was natural that antiquarians should try to discover similar patterns in Britain. The province was continuously occupied from AD 43 to the early fifth century; it had four colonies and a network of roads, many of them conspicuously straight. So it would seem natural for Britain to have had some kind of centuriation. Unfortunately some of the early attempts to prove this were totally unsound. Thus a stone, now lost, found at Manchester[1] recorded that the century of Candidus, ie the company of soldiers commanded by a centurion of that name, built 24ft of wall round the *castellum*. This was absurdly interpreted by H. C. Coote[2] as if it denoted a land holding in a 'century', 20ft in one direction and 4ft in another! Again, W. T. Watkin[3] thought he had found remains in Lancashire and Cheshire of earthen mounds called *botontini*, used as centuriation marks, and in this he was followed by Sir Montague Sharpe and others. But Mommsen's view was that *botontini* were a local feature of North Africa.

On the other hand, the well-known writer on ancient town-planning, Haverfield,[4] in criticising the suggestions made by Sharpe,[5]

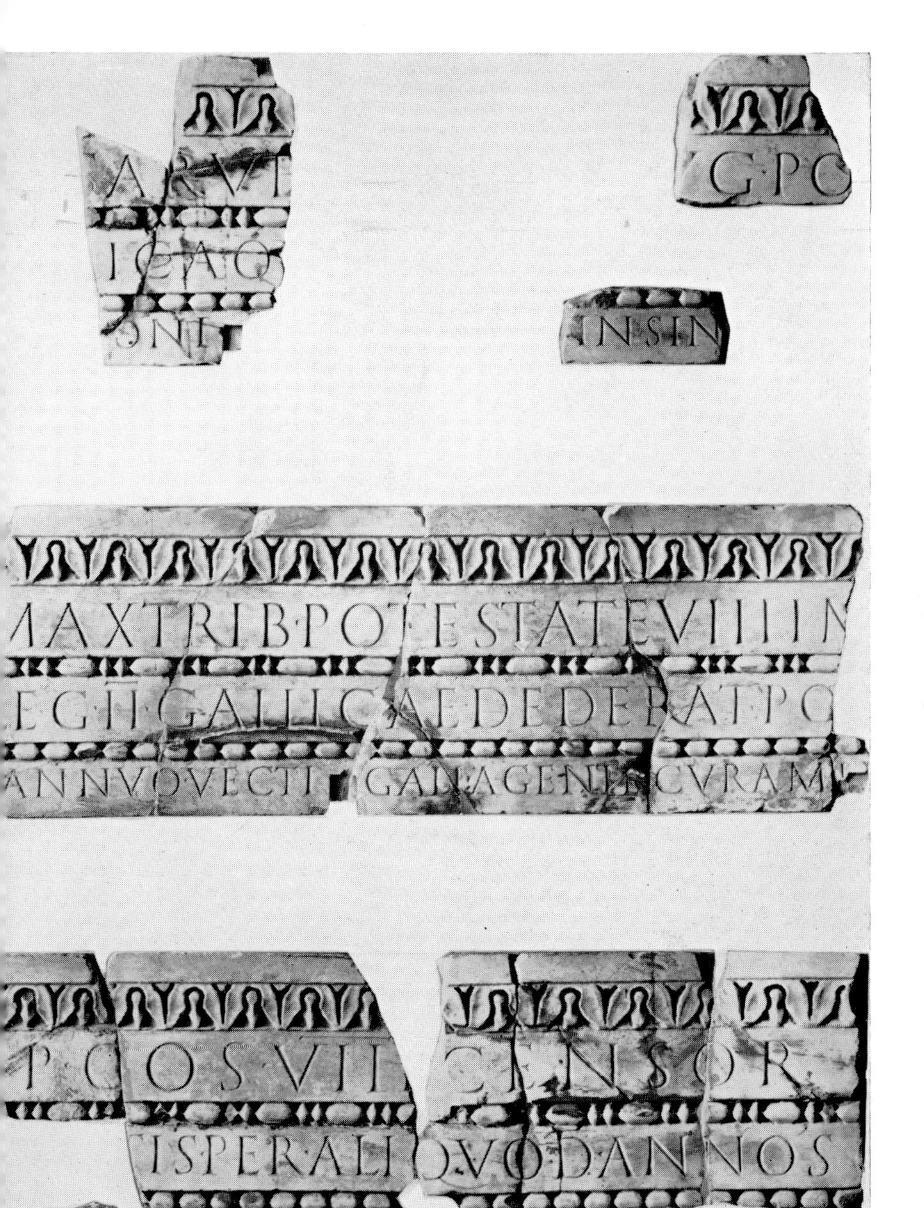

e 189 Inscription of Arausio (Orange), AD 77, setting up a cadaster by order of Vespasian. The ored and expanded Latin text reads: Imperator Caesar Vespasianus Augustus, pontifex maxi-s, tribunicia potestate VIII, imperator XVIII, pater patriae, consul VIII, censor, ad restituenda lica quae divus Augustus militibus legionis II Gallicae dederat, possessa a privatis per aliquod aliquot) annos, formam proponi iussit, adnotato in singulas centurias annuo vectigali, agente curam. . . . Ummidio Basso proconsule provinciae Narbonensis. Translation on p 160

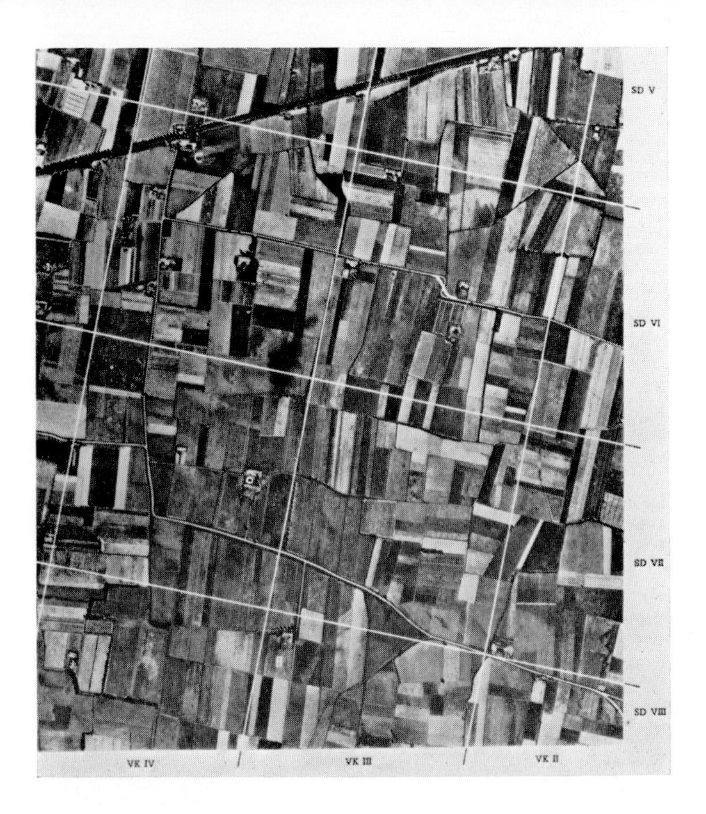

Page 190 (*above*) Air photograph of part of the Rhône valley between Lamotte du Rhône
Notre Dame des Plans, with reconstruction by the late Professor A. Piganiol of the centuriatio
terms of Orange Cadaster B, beyond the *kardo maximus* and to left of the *decumanus maxin*
(*below*) air photograph of the area between Hedjez el Bab, Oudna and Grombalia, Tunisia, show
several patterns of centuriation superimposed. Cliché IGN I

could find nothing better to produce as evidence of centuriation in Britain than two parallel roads near Braintree about 12km apart. The distance can be made to work out at 340 *actus*, but such a multiple is most unlikely, with or without intermediate *limites*.

As to Sharpe's scheme, he started off on the wrong foot by being ignorant of Latin. Thus we are surprised to read:[6] 'A *possessa* was the square block of land contained within four quintarial limites. . . . Its superficial area measured 1,300 *iugera*.' This, of course, is nonsense; there is no such noun as a *possessa*. The whole scheme is based on a total misunderstanding of Hyginus, *De limitibus*:[7] 'legimus in quibusdam locis ab uno mille et trecenta iugera fuisse possessa', 'we have read that in some places 1,300 *iugera* were occupied (with squatter's rights) by one man'. Yet such an error dies hard, and we find it repeated in print.[8] Sharpe then proceeded to contrast these 1,300 *iugera*, wrongly conceived as a regular measure, with the extent of a *saltus*, which he took to be 1,250 *iugera*, and provided for the extra space by allowing for *limites*. For this *saltus* he quoted Siculus Flaccus,[9] who says it is 25 'centuries', but in the context of a regular 'century' of 200 *iugera*, since it is only in the following paragraph that he goes on to irregular ones. So Siculus Flaccus thought of a *saltus* in the surveying sense as 25 × 200 = 5,000 *iugera*, and no other writer in the Corpus contradicts him. To correct some other Latinity and points of detail of Sharpe's: (a) *arcae* in the agrimensorial sense are not holes in the ground but boundary marks consisting of hollow squares; (b) there is no such noun as a *devia*; (c) Frontinus, a modest man, did not call any road Via Julia after his own *gens*; (d) a yoke, with which *iugerum* is connected, is not *iuga* but *iugum*; (e) broken sherds are not *testa fusa* but *testa tusa*. Moreover from the planning point of view Haverfield was right in saying that lines roughly parallel or roughly at right angles are not what we should expect of Roman land surveyors: they should be exact. On the other hand, if Sharpe[10] discovered on one road between the upper Colne and the Lea five crossways in succession at intervals of 10 *actus*, this is surely more than a coincidence. We may dismiss the majority of his conclusions and question the existence of centuriation in Middlesex without discarding such evidence as seems interesting.

We have, however, an almost certain relic of centuriation, whose *limites* have partly disappeared and partly become warped. This is north of Durobrivae-Rochester[11] (Fig 50), where an area south of the village

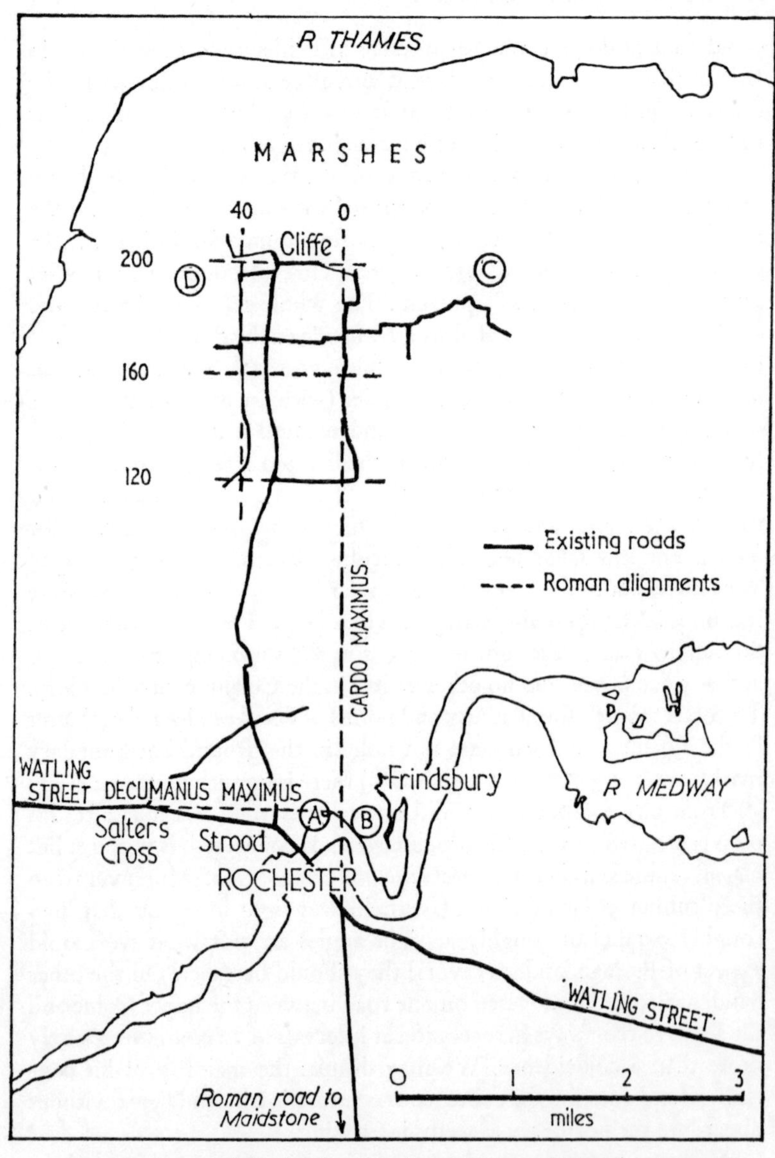

50. Roads near Cliffe, north of Rochester, Kent, which may preserve the pattern of Roman centuriation. Key: (A) Roman road, (B) Roman house, (C) Roman pottery kiln, (D) Romano-British cremation

of Cliffe was examined with the aid of an 1840 tithe map. M. D. Nightingale, the author of the article, supported by C. E. Stevens, argues convincingly that the roads, paths and boundaries in that area are likely to be the remains of a standard centuriation scheme, 20 × 20 *actus*, having its centre at Frindsbury on the left bank of the Medway. The KM is thought to have been a continuation northwards of the Maidstone–Rochester road until it reached the marshy area north of Cliffe, the DM to have been Watling Street, London–Dover, the important road of south-east Britain, which clearly went due east from Salter's Cross.[12] At 120 *actus* north of the Roman Watling Street there are a road and path at right angles, then at 200 *actus* north is the main street of Cliffe, each running east–west. On the 1840 map the KM is represented between 160 and 185 *actus* north by a south–north road, which then turns due east and after 5 *actus* turns north again. One plot measuring 2 × 1 *actus* = 1 *iugerum* belonged in 1840 to a single owner. The phrase *decumanus primus* and the note[13] 'If the *Decumanus Maximus* was Watling Street . . . then this *limes* which I take as the *Decumanus Primus* was 120 *actus* north from it' need qualification. In the first place, although writers in the Corpus are unsure of the correct numbering, the usual view is that the road next to the DM in each direction was *decumanus secundus*, not *primus*, and so on, counting the DM as No 1. Secondly, if the 'centuries' were 20 × 20 *actus*, the *limes* 120 *actus* north of Watling Street must have been D VII, and the intervening ones must have disappeared. Rochester was not a colony, but this in itself is not an obstacle to such a scheme having existed.

A far less orthodox area where there are fairly clear signs of some sort of Roman planning is at Ripe, Sussex, near the Roman road from Lewes to Pevensey[14] (Fig 51). A rectangle of about 3 × 2·5km contains lanes or fields at right angles or parallel to the main road. Although there has again been warping, their distances apart look as if they represent exact numbers of *actus*. East–west along the main road they occur roughly in multiples of 10 *actus*, namely 15, 20, 30, 40, 50, 60, 70, 87. One would imagine that, even though the lane 87 *actus* distant may have been Roman, it did not form one of the *limites*. North from the main road the intervals are hard to define, but south of it they seem to be 6, 12, 18 and 24 *actus*, with a road likely to be Roman but much warped, Langtye Lane, at 24 *actus* distance. The eastern side of the system is represented by the road to Newhaven, which is here at right

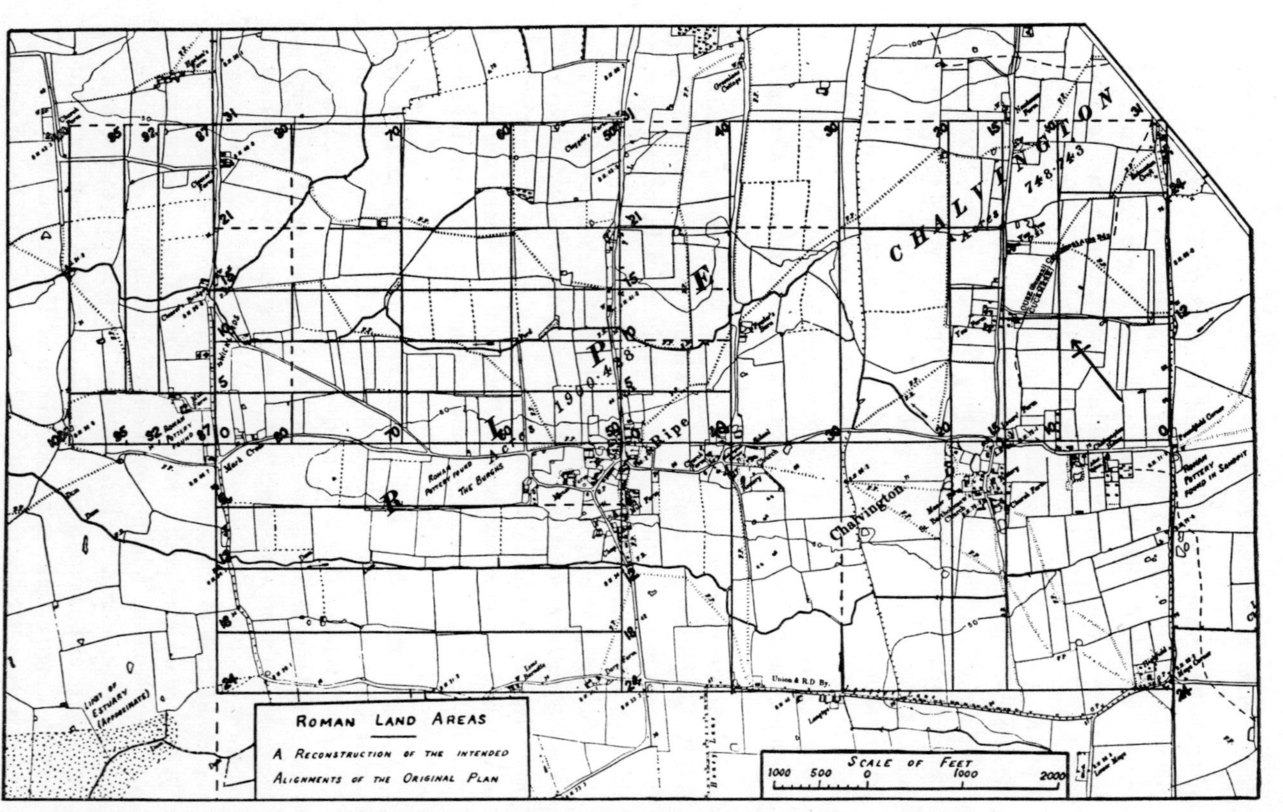

51 Ancient land divisions near Ripe, Sussex. Crown copyright reserved

angles to the Lewes–Pevensey road. There seems in this system to be a confusion between 5 and 6 *actus* units, both of which probably occur in this small area. It does not look like any of the centuriation patterns found in Italy or North Africa, and may not deserve the name. It does, however, look very much like a land assignation made in multiples of *actus* by someone with at least a vague notion of Roman surveying. To what settlement it was attached is uncertain: it is somewhat distant from Anderita (Pevensey).

As to the colonies, these were at Camulodunum, Lindum, Glevum and Eburacum (Colchester, Lincoln, Gloucester and York). No centuriation has as yet been proved at any of them, and the late Sir Ian Richmond argued that at Colchester, founded as the first colony in Britain in AD 49, probably for veterans of the fourteenth and twentieth legions, there was not likely to have been any initial centuriation. The attempt by Coles[15] to show it in a wide area round Colchester is not convincing. He is able to plot squares and rectangles in various orientations, but the measurements do not seem to correspond to likely numbers of *actus*. As to the Gloucester area, Berry[16] thought there might have been centuriation in an 'approximately square' tract between Gloucester and Cheltenham. But the corners of his tract, which have grid references 836195, 898160, 952222 and 889270, show it to be an irregular quadrilateral, not a square or rectangle. Certainly the Cirencester–Gloucester road is straight in this section and led direct to the original legionary fortress at Kingsholme,[17] so that an examination of old maps of the areas north and south of the road might yield results. No centuriation has been traced on the Fylde, where about AD 200 drained land is thought to have been given to Sarmatian veterans stationed at Bremetennacum (Ribchester).[18] Round Newton Kyme, near Tadcaster, rectangular fields adjacent to Roman roads have been examined as possible evidence of allocation *per strigas*.

TOWN PLANS

In general the town plans of Romano-British settlements may be divided into (a) colonies and military forts:[19] perimeter usually rectangular, with rounded corners, roads at right angles, centre often occupied by *principia*; (b) other large settlements, including *civitas* capitals:[20] perimeter usually polygonal and irregular, roads mainly at right angles; (c) smaller settlements: mostly irregular pattern, but with a certain amount of regularity observable. (See M. Todd, 'The Small Towns of Roman Britain', *Britannia* 1 (1970), 114–30.) To these classifications there are, of course, numerous exceptions.

If as a specimen of the first category we study the pattern of Roman York, we find that the military fort lay on the north-east bank of the R. Ouse, in the area near the Minster (recent restoration work on this uncovered parts of the *principia*), while the civilian area was on the opposite side of the river. The streets of the military settlement were not orientated according to the compass points, but were based on (a) the bridge and road to Tadcaster, (b) starting at right angles to it, the road from the centre of the military settlement to Catterick (there was also one from the bridge to this Catterick road, but that did not form an exact right angle), (c) the outer perimeter, which conforms roughly to the direction of flow of the R. Ouse at that point.

For the second type we may take Calleva Atrebatum (Silchester; Fig 52), which was most recently excavated between 1954 and 1958. Here again the interior roads are at right angles, apart from slight deviations, and approximately follow the compass points. But the perimeter consists of an irregular polygon, and the connecting roads do not have the same orientation as the streets in the settlement.

An example of the last type is another Durobrivae, at The Castles, Water Newton, Huntingdonshire. This settlement is just south of the bridge where Ermine Street crossed the R. Nene. The ancient topography can conveniently be studied in the Ordnance Survey *Map of Roman Britain*, 3rd edition, Figure 3. What emerges is the following: (a) the settlement itself, geographically important but small in size, was contained within an irregular elongated hexagon straddling Ermine Street; (b) the roads are mostly in long straight stretches, though the one from the west skirts a loop of the R. Nene; (c) orientated north–south or east–west or approximately so are (1) King Street, leading south from Ancaster, (2) a short stretch of road leading south to Mill Hill, (3) the camp opposite the point where the road from the west meets Ermine Street, (4) the fort north-west of Durobrivae.

ROADS

Britain is one of the best provinces in which to observe the layout of

Roman roads.[21] There is a large and comparatively well-preserved network of them, and many have been examined more carefully than those in other provinces. The main roads mostly radiate out from London, but there are also frontier roads, including the Foss(e) Way,

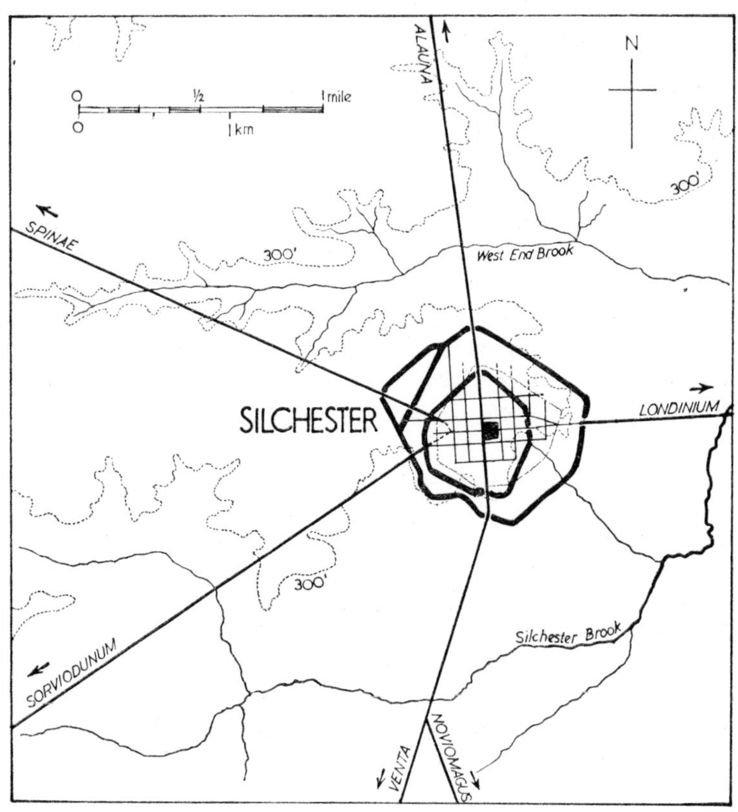

52. Silchester, Hants: outline of Roman town and converging roads

which must have been one such almost at the outset of the Roman conquest, and many connecting roads. The planning of certain roads has already been discussed in the preceding sections.

The immediately noticeable feature is the straightness of the roads where no difficulties of terrain intervened. Many roads have long straight stretches, and the corners, unlike those of military forts, are not rounded off. The result of the straightness is that they often ride roughshod over contours which many more recent roads accommodate, although the motorways tend to go through them with deep cuttings and high embankments. It is obvious that the planning of these straight lines was done on a very large scale. Thus the road from London to Regnum (Chichester), Margary road 15, starts off with a section, as far as Ewell, which points almost directly to Chichester.

Some of the roads present features of parallel planning and right angles. West of the upper Cherwell, Margary roads 561 and 562, Swalcliffe–Warmington and South Newington–Hanwell, run practically parallel. In the area north and west of Gloucester, certain stretches of roads 180 (Gloucester–Worcester), 55 (Cirencester–Kingscote Park, a local road), and 18a (Bourton–Alcester), run on almost parallel courses. Similarly, one can find parallel roads in the Braintree area, as already mentioned, and near Verulamium.

The Foss Way, designed as a frontier road at the time of the Claudian invasion, ran all the way from Axmouth to Lincoln, perhaps with extension to the east coast. It may be thought to have influenced to some extent the orientation of other roads. From Leicester, on the Foss Way, the road ESE to Melbourne and Godmanchester forms a trifle more than a right angle with the section of the Foss Way leading from Leicester to High Cross. One can in fact, as C. E. Stevens pointed out to the author, regard Cirencester–Leicester–Lincoln as an almost straight base line and Leicester–Colchester as the perpendicular to this. If this is not a coincidence, it may possibly indicate planning of the originally conquered part of Britain on a huge scale. But again one may have doubts, since the angles are not exactly 90° and 180°. Also the road in the Cotswolds section was on a very marked geological line of Oolitic Limestone.

THE FENLAND

A survey has just been carried out of the Roman development of the Fenland.[22] This shows that throughout the Roman occupation there is substantial evidence of use of the Fenland by the Romans, but that it was probably imperial land. The absence of regular land divisions on the silts is compared to the region round Padua, where air photographs have shown that the network of centuriation ends where it meets an area of watercourses. Round the Wash there are some large holdings, which do not, however, average out at 200 *iugera*, as one might expect if surveyors had been at work, but at about 175. Around 153230 and 170220 there are long, narrow rectangular fields, only about 60ft broad but 400–600ft long.

There are in the Fenland some noticeable features in rectilinear planning. Many roads have long straight stretches, and three of these exactly or approximately follow the cardinal points, if we include a probably Roman road striking due north until it meets the R. Welland, which it then skirts. Of the waterways, the chief rectilinear one is the Car Dyke, running NNE–SSW with a branch at right angles ESE. In addition, however, to Roman works, there have, in the area of the Wash, been rectilinear reclamation and improvement schemes at various periods since the sixteenth century.

THE ANTONINE WALL DISTANCE SLABS

Both Hadrian's Wall and the Antonine Wall were built by legionaries. In the case of Hadrian's Wall, many inscribed stones dealing with the construction have been found, but the evidence is not very detailed and is difficult to interpret.[23] For the Antonine Wall, on the other hand, we have records of the exact lengths constructed by the various units employed.[24]

This wall was constructed in AD 142 or 143 and ran for 59km from Bridgeness on the Forth to Old Kilpatrick on the Clyde. The legionaries who worked on it, perhaps about 7,000 in all, were of the sixth legion, the twentieth and the second Augusta, stationed at York, Chester and Caerleon respectively, the last-named being present in full strength. The wall was of turf, set for stability on a stone base 4·5–5m wide, with a V-shaped ditch about 6m to the north and a military road 40–50m to the south. There were probably nineteen forts, varying in size, situated at roughly 3–3·5km intervals.

The distance slabs discovered number nineteen; they were set up by units of the three legions to record the lengths of wall completed by them. The western end of the wall, 6·4km long, from Castlehill to Old Kilpatrick, was built in six lengths, each measured in feet. The 53km long central and eastern section, from Castlehill to the Forth, was on the contrary recorded in paces of 5ft. The variation in measurement was at first not appreciated by archaeologists, since P is used as an abbreviation for both *pedes* and *passus*. Theories to account for the discrepancy are that the wall was begun from the east but met unexpected difficulties of terrain, or more probably that the legionaries working near the Kilpatrick Hills were harried by the neighbouring tribes, so that the units had to work in short stretches, in close contact with each other. The Highlands come much closer to the line of the Antonine Wall in the West than in the East, and the West would be regarded as a danger area.

The most prominent distance slabs are those which were erected at each end. The slab from Bridgeness, on the Forth, records that the second legion Augusta completed 4,652 paces. On the right it has a relief showing *suovetaurilia*, a sacrifice of pig, sheep, and bull to mark the inauguration of the wall. On the left it has the figure of a Roman cavalryman crushing Caledonians. The slab from Old Kilpatrick, on the Clyde, records that a detachment of the twentieth legion completed 4,411ft. In the centre of the inscription is a reclining Victory holding a laurel wreath and a palm branch.

It seems highly probable that at least one military surveyor was in charge of each of these important operations. One may only wonder whether the Antonine Wall is not to some extent an example of too many cooks spoiling the broth, since with overall co-ordination even threats by barbarians or unexpected terrain would not have led to two different systems of measurement being used.

14

Roman surveying and today

The object of this final chapter is to try to show that this study is not a dead, academic one but that it has relevance to the modern world and that there is full scope for further research into many aspects of it. With this end in view, some of the preceding chapters will be reviewed to the extent to which they may be relevant.

Roman surveying cannot be understood without a proper examination of its antecedents, and this has been attempted in the second chapter, though only in outline. The ancient Near East provides much opportunity for further excavation. Our knowledge of early civilisations is constantly being pushed further back in time and farther afield; it is now thought that some form of primitive civilisation existed in the Nile valley as early as 13,000 BC. The knowledge of early times can sometimes be correlated with modern techniques; we are, for example, able to appreciate that the idea of a geological map, which was developed in its present form only in the nineteenth century, existed already in embryo form in dynastic Egypt. By comparing finds from different areas we are better able to recognise the exact purpose of each. By way of contrast, when the wooden surveyor's cross was found in the Fayyûm seventy years ago, its function was at first not properly understood.

When we come on to the planning of Greek colonies in southern Italy and Sicily, there is an element of continuity which strikes the observer forcibly. Even if there were throughout the classical period uncouth Romans, like the soldier who in 212 BC killed Archimedes at the siege of Syracuse, Romans were in general keen to learn from Greece both the theory of geometry and to a limited extent its practical applications. Under the Empire surveyors were sometimes Greeks, sometimes of Italian stock. We may take it that Latin, not Greek, was their working language; no corresponding treatises on practical surveying under Roman jurisdiction have been preserved in Greek, and the close connection with Roman law reinforced the retention of Latin. In the Corpus, words dealing with land tenure, boundaries, etc, come from Latin roots, whereas the majority of mathematical terms come from Greek. Legal quotations, apart from an extract from a boundary law attributed to Solon, are in Latin. An interesting parallel is to be found in the anonymous author of the treatise *De rebus bellicis*[1], dating from the fourth century AD and, like the Corpus Agrimensorum, illustrated: there the military technical terms are partly Latin and partly Greek.

There may well still be an urban survival of the Greek system of long narrow strips in towns. Neapolis (Naples), 'new city', was the early successor to a settlement founded about 600 BC as a colony of Cumae, the earliest Greek colony in Italy. The sector of the settlement known as Palaeopolis, 'old city', disappeared when in 327 BC the Romans conquered it. But Neapolis flourished, being long noted as a centre of Greek language and institutions. In ancient times the city occupied an area extending along the coast in what is now the zone round the Cathedral (p 208). It corresponded roughly to an area between S. Giovanni Maggiore and S. Maria del Carmine and inland to about 400m beyond the Cathedral. A great part of this district, particularly round the Cathedral itself, has narrow rectangular blocks about 200m long. They were spotted by Beloch[2] as a possible ancient survival; but when he was writing, it was not possible to appreciate that such a pattern was fairly typical of Greek colonies in southern Italy and Sicily. The length of the blocks presumably corresponded to 600ft, perhaps Aeginetan feet of 33·3cm. A parallel for such urban planning may be seen in the air survey of Heraclea Minoa (Capo Bianco, Sicily).[3] Moreover narrow rectangles seem in at least some Greek colonies to have been a feature of country as well as town planning.

In the third chapter the history of Roman surveying and of the *agrimensores* is traced, to the extent to which our sources permit, from its humble beginnings, through its period of unpopularity with wholesale evictions by the Second Triumvirate, to its heyday in the late first and early second centuries AD, and its rather rigid bureaucratic development under the later Empire. The prominence of Frontinus, as the earliest technical writer in the Corpus, should interest those who know of him in three other capacities, as an efficient governor of Britain and as a writer on Rome's water supply and on military matters. Frontinus was asked if he would like a memorial erected in his honour. 'No,' he replied. 'If I have done my duty, what I have accomplished will be sufficient memorial.' Recent research suggests that the part played by him in the planning of Verulamium may have been underestimated.

The rise of the *agrimensor* as a professional man is linked with his prestige as an arbitrator in land disputes. The influence of Roman law, on which this position as arbitrator rested, not only outlived the fall of the Western Empire but is visible in a number of modern legal systems. Although Cassiodorus obviously liked pulling the legs of surveyors, at the same time he recalled their long history and their importance even in his day.

The main units of measurement used by the *agrimensores* have not survived, though the perch and foot, used for smaller measurements, have left their legacy in various countries, especially in British and American equivalents. The perch, however, has long been obsolescent, and in Britain the foot too is before long to be discarded for the metric system.

Centuriation has left a permanent mark in the countryside of the Po valley, Campania and many other areas. In Tunisia it is traceable over vast expanses of territory, and much more must lie buried under sand. The possibilities for plotting it throughout the Roman Empire have increased enormously with the development of air photography and its interpretation.

It is of interest to compare the Greek rectangles and Roman squares with more recent land divisions in other countries. Japan, with only a restricted amount of land at its disposal, had a regular system of small squares (p 139), usually orientated by the compass points. In Holland, in the first half of the seventeenth century, the Beemster polder west of Edam was divided by roads into squares with sides of approximately one nautical mile.[4] Within these are rectilinear canals, often one-tenth of a mile apart; the straight lines would assist with the drainage of the flat land. The similarity of schemes of 'bonificazione' in Italy to the Roman pattern has already been mentioned.

Two early plans for the division of American lands into squares, before Jefferson's plan, are known. In 1717 Sir Robert Montgomery proposed for South Carolina a system of square districts, twenty miles to a side, partly composed of square-mile cells. In 1765 General H. Bouquet suggested, for the upper Ohio, fortified squares with sides of one mile, each having five adjacent squares of the same area for arable land, pasture and woods.

The plan submitted by Jefferson in 1784 provided for a land division by the compass points, by squares to be known as hundreds, with sides of ten geographical miles of 6,086·4ft. These hundreds, 100 geographical square miles in area, were to be divided into lots of one square mile = 850·4 acres each. The details of the plan were worked out by Hugh Williamson, a native of Pennsylvania, who from 1760 to 1763 had been a professor of mathematics at the College of Philadelphia.[5] It is rightly claimed by Pattison that his name should be coupled with that of Jefferson as joint authors of the plan. It seems very likely that this Jefferson-Williamson plan goes back to the seventeenth-century Dutch land division. Both Bouquet, whose book Jefferson possessed, and Williamson, whom he consulted during the winter of 1783–4, had connections with Holland.

The question has several times been raised whether Jefferson's ideas go back to the Roman land surveyors. If they do, the connection could be roughly as follows. Hugh Williamson, his collaborator, was a Doctor of Medicine of Utrecht and a member of the Holland Society of Sciences and the Society of Arts and Sciences of Utrecht.[6] In 1674 Willem Goes had published in Amsterdam an edition of the Agrimensores, with diagrams (many inverted), under the title of *Rei agrariae auctores legesque variae. . . .* This was the third illustrated edition, the first two having been those of Gallandius and Turnebus, Paris, 1554, and of Rigaltius, Paris, 1614, the latter with only a few diagrams. It is very probable that Goes's edition would be well known to a mathematician, interested in astronomy, such as was Williamson. If the origin of the one-mile squares is to be sought in a Dutch polder, at least the orientation by the compass points, and perhaps the basic idea, may certainly go back to the Corpus Agrimensorum. The polder division may possibly have been inspired by relics of ancient squares such as exist in Limburg, though it is very doubtful if these were noticed before the present century.

Another point, the Roman unit *centuria*, has been left unmentioned, perhaps on purpose, by writers on the beginnings of the American survey system. It suited Jefferson's decimal scheme, which led him at the same time to decimalise the coinage, to incorporate the term 'hundred'. In English land usage this had at first meant 100 hides, a hide representing a holding of 48 or 120 acres according to the district. In course of time the hundred lost its 100-hide equation and came to mean simply a subdivision of a county with its own court. In this sense it was adopted also in the colonies of Delaware, Maryland, and Virginia. In Virginia, however, it was not an effective division, though Jefferson was always trying to make it so, as an equivalent of the New England township. But his proposals had included an educational bill of 1778 in which he advocated bounding hundreds 'by watercourses, mountains, or limits'. On the other hand the Latin word *centuria*, where applied to land, always denoted a square or rectangle. In the great majority of the schemes it was an exact square, though admittedly much smaller, usually 50·4ha. In the present work it has been called 'a "century" '; but since it meant 100 *heredia*, if one wanted an English equivalent having a related significance, which 'century' does not normally, one might well think of 'a hundred'. To a lover of decimalisation the difference in size would hardly matter. What would not appeal to Jefferson would be such Roman measures as the *actus*, based on the duodecimal system, and the *iugerum*, which at 200 to the square plot would spoil the decimalisation. But he and Williamson, even if they looked primarily to the Anglo-American hundred, may have had the *centuria* too at the back of their minds in plotting square hundreds.

The American survey was confronted with a problem which did not worry the Romans. As a letter from Timothy Pickering in 1785 pointed out, each hundred was to have sides of 60,864ft, yet the east and west boundaries were to be true meridians; and meridians, as Jefferson evidently knew but had not provided for in legislating, converge. This dilemma was solved by a system of base lines 25 miles apart (Fig 53), at which the north–south boundaries were offset. The Land Ordinance of 1785 modified Jefferson's proposals. The system of squares and the orientation by compass points were retained; but squares of sides six statute miles long, each with a township, took the place of his squares with sides of ten geographical miles. In Canada a rectangular system orientated by the compass points was adopted, each rectangle having a length twice that of its breadth.

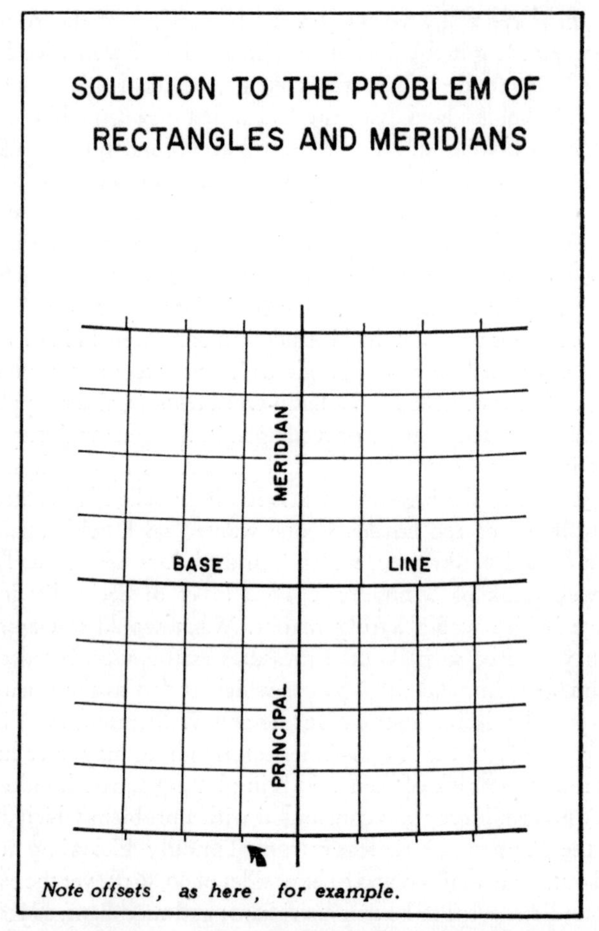

53. How the USA surveyors solved the problem of converging meridians

In the field of the history of cartography the Corpus Agrimensorum has a contribution which is not always fully recognised. Admittedly the extant MS miniatures are from the sixth century AD onwards, but although they are often corrupt, their general impression, well-preserved colouring and many details no doubt reflect those of the originals. These gave a necessary visual aid to the learner, and if we may judge by their copies show some skill in cartographic execution. Fields,

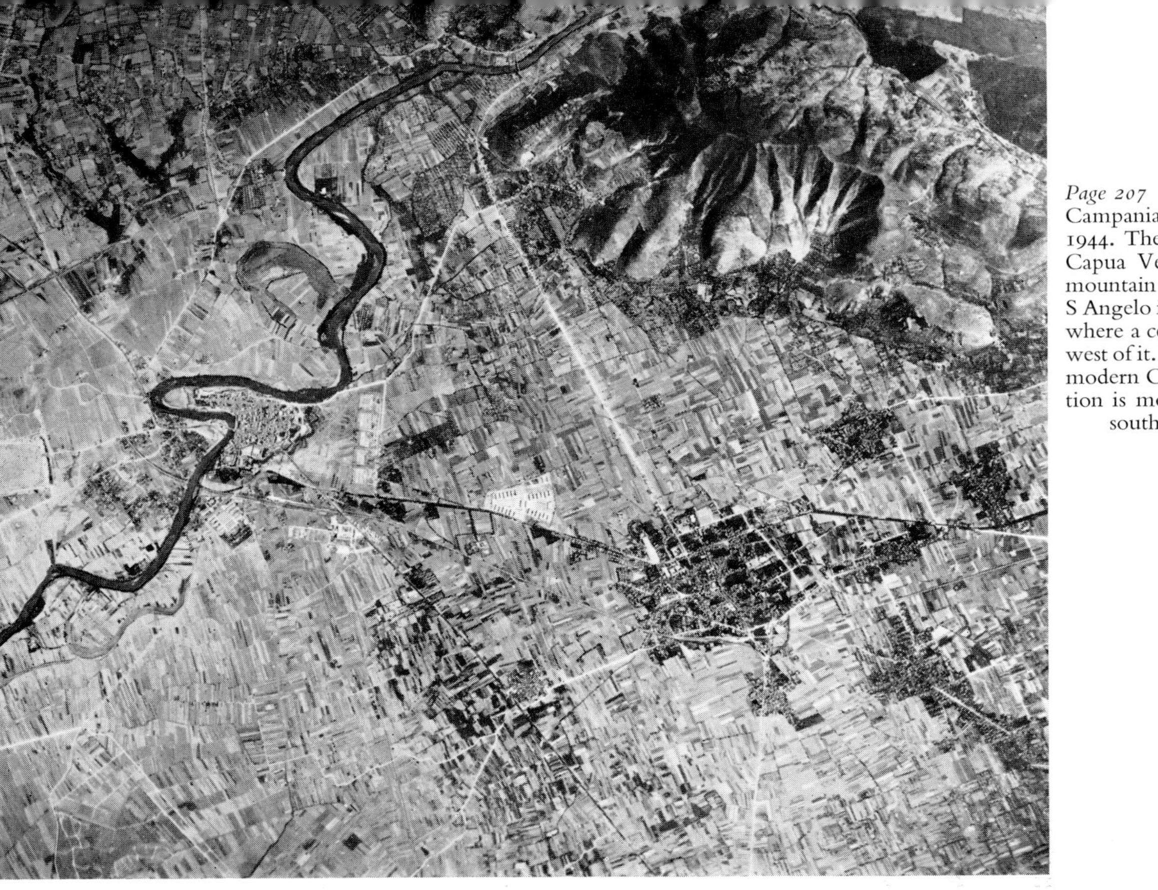

Page 207 Air photograph of part of the Campania plain looking NNE, taken in 1944. The largest settlement is S Maria Capua Vetere, the ancient Capua. The mountain NNE of it is Mt Tifata, with S Angelo in Formis (Latin *forma* = map), where a centuriation stone was found, to west of it. The river is the Volturno, with modern Capua in a meander. Centuriation is most easily seen north, east and south of S Maria Capua Vetere

Page 208 Air photograph of central Naples, taken in 1944. This illustrates the survival of a rec angular street pattern. Top left, part of Museo Nazionale; bottom right, S Annunziata; cent right, Castel Capuano; slightly to right of middle, the Cathedral

mountains, trees, buildings, and walled cities, the last somewhat symbolically represented, are all drawn or painted carefully and with individual features. It is typical of ancient illustration that, although there is little regard for perspective or comparative size, details of buildings such as Corinthian or Ionic capitals are often drawn in with great care. We can profitably compare the topographical drawings with the even smaller ones in the Peutinger Table and with comparable material illustrated in the recent book on it.[7]

On the Corpus itself much research is needed. It is always just possible that another manuscript of at least part may turn up. Where a treatise has lost its heading, the truncated text is sometimes wrongly attributed and misleadingly entered in a catalogue. But if this is only a distant hope, there is a large body of material to hand, to which justice has been done from the point of view of MS tradition, but which has never been translated except in part, and on which there is no continuous commentary. The corruption of the text and unintelligible jargon make this difficult, but it is worth attempting for several of the authors. The essays in Blume,[8] good as they are, need amplifying with more recently available material. It would be helpful if the Vatican MS as well as the Arcerianus could be reproduced in facsimile, preferably with colour, since much is lost in black-and-white reproduction of the miniatures, especially when the reproductions are small.[9]

The study of centuriation is the aspect on which most research is being carried out and on which most is being written today. A good introduction to this type of work was published by Castagnoli,[10] particularly valuable for Italy. But a detailed critical bibliography is needed, especially as many of the articles on limited areas of centuriation are in local periodicals not always easily accessible in other countries. Italy, because of the diversity of its local allocations and the amount of historical information we possess about it, continues to occupy the most important place in topographical research into land division. The centuriation map on which General Giulio Schmiedt and others are working should encourage further studies. It might have been hoped that the *Atlas des centuriations romaines de Tunisie* would have led to extensive new research; but the similarity and regularity of the centuriation makes for a lack of variety, and there are problems of accessibility and resources. Despite the unfortunate destruction of much relevant epigraphical material at Orange, there is still scope for further work, particularly in the preparation of a more complete aerial survey and in attempting to equate this with the evidence to be obtained from Piganiol,[11] perhaps with the assistance of old maps. If anything similar to the Orange cadaster stones turns up in future, probably the first task will be to feed all the information received into a computer.

If more Roman surveying instruments are found, we may be able to determine whether the simple Pfünz type or the more sophisticated Pompeii *groma* was the one normally used. It would also be interesting if the metal parts of a *chorobates* turned up. It has been claimed, but on what evidence is doubtful, that centuries after the fall of Rome some instruments and methods survived, and that they were improved in the fourteenth and fifteenth centuries by European instrument makers. The perch, Latin *pertica*, which had been among other things a 10ft surveyor's rod in Roman times, survived in the Middle Ages, but with varying measurements. In the late Saxon and early English periods, whereas in the south of England it was 12ft, in the north it was as much as 22ft. By Tudor times a different type of instrument, the cross-staff, and a different procedure, triangulation, had evolved in England.[12]

The study of Roman colonies has made considerable progress during the past fifty years.[11] As the importance of colonies in the power and prestige of Rome has become increasingly recognised, much research has been carried out, some on colonies in general, some on individual areas or periods. In certain areas historical and geographical studies have been combined, the research including a detailed investigation of centuriated land. Air photography of the more fertile lands round all colonies would be most desirable. At Emerita (Mérida) the public buildings are so extensive and so well preserved that we can easily form an image of the importance of this Augustan colony. The Corpus gives us details of its centuriation and of the legal position of land by the R. Ana (Guadiana). Yet observation in the field and study of maps have as yet revealed nothing—surely a case where air photographs taken under ideal conditions and interpreted by an expert might yield a positive, even if limited, result. In North Africa, if the difficulties of continuous research can be overcome, more detailed investigation both of colonies and of centuriation should prove rewarding.

Whether the same would apply to Britain is far more doubtful. Coverage of archaeological sites since World War II has been fairly extensive. It seems likely that in this remote province there was only comparatively little regular, rectilinear land division. But a study of early local maps of areas round the most prominent Roman settlements might show more parallel roads, paths, and boundaries, even if warped, than can be found on modern maps. These should not, however, be accepted as Roman unless they can be shown to be a likely number of *actus* apart. In addition, however, to possible centuriation, work can be done on the plotting of Roman roads in Britain. Not only have many never fallen out of use, but in many places recently constructed motorways follow the lines of Roman roads. The accurately planned straight stretches of these undoubtedly look like the work of surveyors, but we cannot prove this.

What today seem advanced techniques for the discovery and recording of ancient land divisions may in future be regarded as to some extent obsolete. With improvements in telescopic photography, the ideal method may prove to be satellites continuously photographing the whole world for every required purpose. Computers would then confine the areas to be examined for centuriation to likely places in the Roman Empire. After this they would be programmed to reject photographs which had no features of ancient land division. Finally they would classify and index every area of any significance.

The Romans were not great scientific inventors; their education and their outlook on life did not run on those lines. But they were determined and methodical, and their land surveys, carried out over hundreds of years often on a gigantic scale, had a profound influence not only on the whole world of antiquity but indirectly on the lands of the Roman Empire, and perhaps elsewhere, ever since. Their motto could well be: *si monumentum requiris, circumspice.*

Finding list

MANUSCRIPTS OF THE CORPUS AGRIMENSORUM

(a) *Fully illustrated.* The early representatives, up to ninth century, are:

Arcerianus A, 36, 23 Aug 2°, Herzog August Bibliothek, Wolfenbüttel. For facsimile reproduction, see bibliography to Chapter 9, Codices.

Palatinus Vaticanus Latinus 1564, Biblioteca Apostolica Vaticana, Vatican City.

(b) *Illustrated with geometrical drawings only.* The early representative is Laurentianus Plut XIX cod 32, Biblioteca Laurenziana, Florence.

(c) *Not illustrated.* The early representative is Arcerianus B, 36,23 Aug 2°, Herzog August Bibliothek, Wolfenbüttel.

INSCRIBED CADASTER STONES

Musée Archéologique, 84 Orange (Vaucluse). Published in Piganiol (1962); many subsequently destroyed in a collapse, but the most important surviving ones now affixed to walls of ground-floor hall.

FORMA URBIS ROMAE

Musei Capitolini, Rome; affixed to wall of courtyard.

PEUTINGER TABLE

Staatsbibliothek, Vienna.

SURVEYING INSTRUMENTS

Contents of Pompeii workshop: Museo Nazionale Archeologico, Naples (the *groma* is in a locked technical hall). Models of reconstructed groma in Science Museum, London SW7, and elsewhere. Merkhet: Staatliche Museen, East Berlin, C2, Bodestrasse 1-3 (copy in Science Museum, London). Fayyûm instrument: Surveying Gallery, Science Museum, London.

Sundials: a good number of Greek and Roman ones preserved, including seven in Museo Archeologico, Aquileia (Udine). A Roman portable sundial: Oxford Museum of the History of Science, Broad Street, Oxford.

MUSEUM OF ROMAN CIVILISATION

Museo della Civiltà Romana, Piazza Giovanni Agnelli, EUR, near Rome. The museum, open from 9 to 14 hrs with later opening some days, is at the end of Viale della Civiltà Romana, which leads east from the Piazza Italia in the EUR, a museum and sports area near the end of Via Cristoforo Colombo, about 7km SSW of the Coliseum. It has very good sections dealing with surveying, roads, bridges, military technology, etc.

CENTURIATION AREAS

For a summary of these see Chapter 10. Among many others may be mentioned:

A large part of the Po valley.
Campania, round S. Maria Capua Vetere and to the south.
Terracina, to examine a small area illustrated in the Corpus.
Pula (Pola).
Tunisia, especially round Carthage and other areas of the North.

Glossary

Actus 120 Roman feet; literally 'driving'
Actus quadratus Square *actus*, 14,400 square Roman feet
Agentes in rebus Under late Empire, corps of imperial agents; literally 'those acting in affairs'
Ager Field; land
Ager arcifinius Frontier territory; unsurveyed land
Ager publicus State land; see Chapter 12
Agrimensor Land surveyor; literally 'measurer of land'
Ara compitalis Altar at crossroads
Arca Hollow, square boundary mark; literally 'box', 'chest'
Arcifinius See *ager*
Area Public or sacred area of town
Aroura In Ptolemaic Egypt, 100 cubits square; literally 'arable land' (Gk)
As, plural *asses* Bronze coin; weight
Asteriskos Simple surveying instrument; literally 'little star' (Gk)
Bes $\frac{2}{3}$ *iugerum*; $\frac{2}{3}$ *as*
Bonificazione System of land reclamation and division in Italian states and modern Italy
Botontinus Earth mound used as landmark
Cadaster Large-scale land survey carried out for taxation
Cardo See *kardo*
Casae litterarum Treatise listing types of country estate; literally 'houses (earlier, cottages) of letters', since drawings are intertwined with letters of the alphabet
Castellum Fort
Castrametatio Survey of camp (*castra*)
Centuria 'Century' of land, originally 100 *heredia* of 2 *iugera* each, hence usually a square of 200 *iugera*, but other measurements found; see Chapter 6
Centuriatio See *limitatio*

Chorobates Long levelling instrument; literally 'area walker' (Gk)
Civitas Citizenship; state; tribal unit recognised as such by Romans
Codex Book in quire form, as opposed to *volumen*, roll
Collegium Guild
Colonus Settler; tenant farmer; from *colere*, 'to till'
Colony Greek colony (*apoikia*) = maritime trading settlement; Latin colony = settlement with restricted voting and other rights; Roman colony = settlement with full rights, often for legionary veterans; under the Empire, some towns given title of *colonia*
Compitum Crossroads
Conciliabulum Small urban district
Cultellare To level; literally 'cut with knife'
Decempeda 10ft surveyor's rod
Decumanus (*decimanus*) One of the two sets of parallel *limites*, usually east–west; literally '(?) outsize balk'. See Appendix B
Decumanus maximus (DM) One of the two main intersections; literally 'largest (?) outsize balk'
Denarius 10 *asses*, later 16 *asses*
Dioptra Astronomical or surveying instrument; literally 'sighter' (Gk)
Evocatus Volunteer ex-legionary; literally 'called out'
Ferramentum Iron instrument; *groma*
Fines Boundary, boundaries; literally 'ends'
Forma Map, plan; literally 'shape'
Geometres Measurer of land (Gk); geometer
Groma Surveying instrument; see Chapter 5
Gromatici Late name for *agrimensores*
Heredium Heritable plot; 2 *iugera*
Insula Island; block of buildings
Iter Journey; roadway; itinerary (also *itinerarium*)
Iugerum 28,800 square Roman feet, 0·252ha; literally, land that could be ploughed in one day
Jóri Japanese system of land division
Kalendarium Public ledger of debts, which were due on kalends
Kardo (*cardo*) One of the two sets of parallel *limites*, usually north–south; literally 'hinge'. See Appendix B
Kardo maximus (KM) One of the two main intersections; literally 'largest hinge'
Komatogrammateus Village clerk (Gk) in Ptolemaic Egypt

Kuduru Babylonian votive stone

Latifundium Large estate

Latini Iuniani A class of freedmen; literally 'Latins under the Lex Iunia'. See p 39

Libella One-tenth of *denarius*; plumb-line level

Libra Scales; pound weight; plumb-line level

Libri aeris Mapping registers; literally 'books of the bronze (map)', *aes*

Libri coloniarum Two treatises listing colonies, with comments

Limes, plural *limites* Road, track or path forming a boundary between 'centuries'; literally 'balk', 'dividing strip'. Also, in singular, area near frontier of the Roman Empire

Limitatio Centuriation, dividing of land by intersecting *limites*

Magister officiorum Head of civil service under late Empire; literally 'master of duties'

Mensor Measurer: (a) land surveyor, (b) military surveyor, (c) architectural surveyor, (d) corn measurer

Meris, plural *merides* Stand for booth; literally 'lot' (Gk)

Merkhet Egyptian surveying and astronomical instrument

Meta Conical turning post; surveyor's mark

Metator Land surveyor or military surveyor

Municipium 'Borough' is the nearest equivalent; see *Oxf Class Dict*

Norma Carpenter's square

Pagus Village; rural district

Passus Pace (double), 5 Roman feet; *mille passus* = 1,000 paces, 1 Roman mile; *duo milia passuum* = 2 Roman miles

Pertica 10ft surveyor's rod; whole of a centuriated area

Pes, foot, and *pes Drusianus*, 'foot of Drusus' scale'; see p 82

Podismus Measuring by feet; measuring in general

Possessor Occupier (with squatter's rights, *possessio*)

Praefectura Assize town; office of *praefectus*

Primicerius Chief surveyor under late Empire; literally 'first on wax tablet'

Principia (plural) Headquarters building of camp

Quintarius Every fifth intersection, wider than others

RP = *rei publicae*, 'belonging to the State'

Rigor Straight-line boundary with no width; literally 'rigidity'

Saltus Grove; 5,000 *iugera*

Scamnum Rectangle of land bounded breadthwise; literally 'bench', 'balk'

Schoenion Rope, measuring-line (Gk); 100 cubits

Scripulum (*scrupulum*) 100 square Roman feet; one twenty-fourth of an ounce or of an hour; literally 'little stone'

Semuncia Half *uncia*

Sevir Man honoured by local community in Italy; literally 'member of board of six'

Sextula One-sixth of *uncia*

Singula One-tenth of *denarius*

Stella Simple surveying instrument; literally 'star'

Striga Rectangle of land bounded lengthwise; literally 'swath', 'windrow'

Subruncivus Narrow *limes*; farm-track; literally one that is weeded

Subsecivum Remnant of land; literally 'part cut under'. See p 94

Suovetaurilia (plural) Sacrifice of pig, sheep, and bull

Tabulae aeris Records on bronze (*aes*)

Tabularium Record office, in which *tabulae*, tablets, and other records were kept

Terminus Boundary mark

Territorium Land falling under jurisdiction of a local community

Terruncius One-fortieth of *denarius*

Tetrans Main intersection; literally 'quarter', 'quadrant'

Togatus Augusti Honorific title under late Empire; literally 'toga-wearing Emperor's man'

Tresviri Commission of three (p 35)

Triens One-third of *iugerum*

Uncia Ounce; 2,400 square Roman feet

Varatio Diagonal sighting; from *varus*, 'crooked'

Versus 100ft (Oscan and Umbrian); literally 'turn', 'row', 'line'

Via publica Road owned by local authority

Vir clarissimus (VC) Honorific title under late Empire; literally 'most famous man'

Notes and references

Chapter 1 Introducing the *agrimensores* Pages 15–18

1 For a short summary see G. D. B. Jones in *Oxford Classical Dictionary*, 2nd ed, Oxford, 1970, s.v. Centuriation; for details, see Chapters 3–12

Chapter 2 Pre-Roman surveying and geodesy Pages 19–30

1 L. W. King, iii (1900), 25. See Bibliography
2 Gadd (1965)
3 (1957), 35
4 Turin, Museo Egizio; J. Ball, *Egypt in the Classical Geographers*, 180–2 and Plates VII and VIII
5 2. 109
5a Borchardt, 1899
6 *Odes* 1.18.1–2. It has been supposed that Horace confused him with Archimedes, who wrote a treatise on the theory of counting sand particles in astronomical numbers; but R. G. Nisbet and Margaret Hubbard in their commentary on *Odes* 1 (Oxford, 1970) discount this
7 Wycherley, 1949; Martin, 1956
8 Castagnoli, 1956
9 Guy, 1964, 122–4. For Metapontum see Schmiedt and Chevallier, 1959
10 *Laws* 741C, 745A
11 Westerman, 1921
12 Tebtunis Papyri I, Appendix I
13 Ox Pap ed Grenfell and Hunt, VI; Déléage, 1934, 132
14 VI. 279. 12
15 1934, 98
16 Westerman, 1921

Chapter 3 Roman land surveying and surveyors Pages 31–46

1 *Aeneid* 5.755–6
2 *Metamorphoses* 1.136
3 *Corp Inscr Lat* X.3825 *iussu imp Caesaris qua aratrum ductum est*

4 *Fasti* 4.825
5 1909, 15
6 1.44.3
7 *De lingua latina* 7.7
8 Nissen, 1869
9 G. Körte, 'Die Bronzeleber von Piacenza,' *Röm Mitteil* 20 (1905), 348–79
10 Heurgon, 1959; Pfiffig, 1961
11 M. Pallottino, *Etruscologia*, 6th edn, Milan, 1968, Pls 18, 21
12 Salmon, 1969, 167, n16
13 1954, 225–6, n4
14 MacKendrick, 1954, 225
15 An accepted emendation for *ianitores* in Cicero, *De lege agraria* 2.32
16 Nicolet, 1970
17 *Corp Inscr Lat* I^2 584; Warmington, 1967, 262–71
18 *Corp Inscr Lat* I^2 585; Warmington, 1967, 370–437
19 14.10 and 11.12
20 Blume i.395
21 Blume i.239
22 *Corp Inscr Lat* V 6786
23 A. M. Duff, *Freedmen in the Early Roman Empire*, 2nd edn, Oxford, 1958; Duff discusses among other things the standing of *seviri*
24 Pauly-Wissowa, *RE* s.v. Gromatici, col 1891
25 Digest: Ulpian, fr 1
26 Blume i.54
27 *Corp Inscr Lat* X 8038
28 *Silvae* 4.3
29 *Epist* 10.17B
30 Blume i.121
31 Dessau, *Inscr Lat Sel* 5947a
32 *Corp Inscr Lat* III 567
33 Details in Blume ii. 248–50
34 A. D. E. Cameron, *Claudian: Poetry and Propaganda at the Court of Honorius*, Oxford, 1970; more briefly, O. A. W. Dilke, *Claudian. Poet of Declining Empire and Morals* (1969), 3–5
35 *Variae* 3.52
36 *Monumenta Germaniae Hist*, *Epist* I.1, *Gregorii I Registri Lib I–IV*, ed P. Ewald (Berlin, 1887), 484–5, Letter VII.36

Chapter 4 The training of Roman land surveyors Pages 47–65

1 *Ars Poetica* 325ff
2 1. 3–10

3 Blume i.166
4 Blume i. 354–6; see now K. D. White, *Roman Farming*, London, 1971
5 5. 1–2
6 Columella ii (1954), 8–9
7 Abstract of article in *Nature* 88 (1911–12), 158
8 Blume *et al.* i, Fig 342
9 Cantor, 1875
10 For a discussion of Diophantus' date see Neugebauer (1957) 178–9
11 Nissen, 1906, I. 79ff; Rose, 1923
12 1.6.6–7
13 Mollweide (1813) 398–9, quoted by Cantor (1875) 68–9
14 Given in Dilke (1967) 18
15 Lewis and Short's *Latin Dictionary* wrongly gives 'to level land with the coulter'
16 Blume i. 33–4
17 *Varo* does not, as Lewis and Short's *Latin Dictionary* gives, mean 'to bend, curve' (the adjective *varus* means among other things 'knock-kneed'), but something like 'to take a diagonal'
18 Hyginus, 1848, 149–51 and Pl i, Fig 9
19 *Georgics* 1.233ff
20 3. 247–8
21 See Boll and Gundel (1924–37), Tannery (1893), van der Waerden (1951)
22 Weber, 1891
23 1891, 26
24 Blume i.45
25 Blume i.139
26 Blume i.36
27 Blume i.149

Chapter 5 Roman surveying instruments Pages 66–81

1 See p 67; Schöne, 1901, 129
2 Della Corte, 1922; Walters, 1921–2; Shorey, 1926–8; Lyons, 1927
3 *N.H.* 7.213
4 9.7
5 Kenner, 1880, 9–20
6 Diels, 1920, 185–92; Singer, 1923
7 Stebbins, 1958
8 Price, 1969
9 Nowotny, 1925
10 8.5.1
11 Needham, 1954–62, iii.218

12 10.9
13 Singer, 1923; Shorey, 1926–8

Chapter 6 Measurement and allocation of land Pages 82–97

1 Castagnoli, 1956; Fletcher, 1967
2 *N.H.* 18.49
3 5.1.7
4 Blume i.159
5 Castagnoli (1964) gives 23 × 12
6 Rose, 1923
7 Bradford (1957), pp 150 and 178–83, with map p 181
8 Lex Mamilia 55. Contrasts and parallels between colonies and *municipia* may be seen in Salmon (1969), index s.v. *municipium* compared with *colonia*
9 References in Dilke, 1967, pp 21–2
10 Castagnoli, 1964
11 Saumagne, 1928
12 Saumagne, 1952, 299
13 Castagnoli, 1958, 27–9
14 Discussed in Dilke (1967), p 14
15 *Corp Inscr Lat* IX 5420: the emperor's ruling on the area ends *possessorum ius confirmo*, 'I confirm the rights of the occupiers (of *subseciva*?)'. For *subseciva* see also Rudorff in Blume (1848–52), ii.390–3
16 Castagnoli, 1956
17 Salmon (1969) Fig 3
18 Blume i.206
19 Blume i.122–3, ii.421. The MSS give different totals, and have been amended by the editors to suit what seems to be the approximate total (in Blume ii.421, last line, 1302⅓ is a misprint for 1356⅓)
20 Blume i.113 and 199–201
21 The MSS differ, and the reading of Thulin (1913, 164) is adopted

Chapter 7 Boundaries Pages 98–108

1 Mommsen (1892)
2 Rudorff in Blume ii.242 seems wrong in thinking that the stone thrown down, with an imprecation to Jupiter to throw out the speaker if he is knowingly cheating, is a boundary stone
3 Blume i.141
4 Blume i.302
5 For a summary of Roman legislation on the subject, see Pauly-Wissowa s.v. *Terminus motus*

6 Rudolph, 1935, pp 224ff
7 Thomsen, 1947
8 Blume i.114
9 Quoted in Blume ii.257–9
10 Warmington (1967), 170–97
11 *Tablettes Albertini* ed Courtois *et al*, 1952; Väänänen, 1965
12 Dilke (1967), Pl 7, Fig 30; translation on p 20 of article
13 Blume i.340, 353, 357–8
14 In Blume ii.223–6
15 A. Rudorff in Blume ii (1852), 422–64
16 ibid, p 423
17 *Fasti* ii, 675ff

Chapter 8 Maps and mapping Pages 109–25

1 *N.H.* 3.17
2 L. Bagrow, 'The Origen of Ptolemy's Geographia', *Geografiska Annaler* 1943, 318–87, argued that the Geography in eight books as we have it is a Byzantine compilation
3 Miller, 1962; Levi, 1967
4 ed Seeck, 1876
5 Richmond and Crawford (1949)
6 Carettoni *et al*, 1960
7 For a smaller plan we may compare *Corp Inscr Lat* vi.1261, which has a diagram of some Roman waterworks
8 An inscription from a geographical monument which included a bronze map is recorded by H. Bazin, *Revue d'Archéologie* 2 (1887), 325
9 Blume i.196
10 Dilke, O. A. W. and M. S., 1961
11 8.21.11
12 J. Johnson, *Excavations at Minturnae*, 2 vols, Rome and Philadelphia, 1933–5
13 *Possessio* does not necessarily in Roman law imply ownership; see E. G. Hardy, *Roman Laws and Charters* (Oxford, 1912), p 36
14 B. W. Frier, 'Points on the Topography of Minturnae', *Historia* 18 (1969), 510–12, with references
15 Schulten, 1898
16 Salmon, 1969, pp 133–41
17 Grosjean, 1956
18 For the treatment of perspective in ancient painting, see M. Pirenne, *Optics, Painting and Photography* (Cambridge, 1970)
19 Dilke, 1961a, p 424

Chapter 9 Roman surveying manuals Pages 126–32

1 The title *gromatici* (from *groma*) has tended recently to be discarded as being late Latin. This was the form adopted for a Latin inscription in the first building of the Ordnance Survey
2 Sir Brynmor Jones at the inauguration of the University of Leeds Television Centre
3 Beeson, 1928; Lowe, 1959, 39
4 Blume ii.10
5 Byvanck, 1923
6 Blume (1848); the diagrams were organised by Lachmann
7 *Imago Mundi*, 1967
8 1.1.4

Chapter 10 Centuriation Pages 133–58

1 Thulin, 1909
2 *Corp Inscr Lat* III 3224
3 Falbe, 1833, 54ff
4 Kandler, 1855, 178
5 Mancini *et al*, 1957, 249–51
6 Nègre, 1963, 65
7 Campana, 1941
8 Tanioka, 1959; Castagnoli, 1958, 30
9 Hawkes, 1947; Richmond, 1969, 135ff
10 Seel, 1963
11 Stevens, 1937
12 Bradford, 1957, 29–34
13 Castagnoli, 1958
14 *Corp Inscr Lat* I² 640
15 *Corp Inscr Lat* I² 2678–2708 and Staedler, 1942
16 Castagnoli, 1953–5
17 Bradford, 1949
18 Mertens, 1958, 1969
19 Castagnoli, 1956; Salmon, 1969, n47
20 Chevallier, 1960, 1088
21 42.4.3
22 Chevallier, 1960, 1086
23 Casteggio; *Athenaeum* 1958, 117
24 39.55
25 Pilla and Bosio, 1965–6
26 Degrassi, 1954, 18–20; Bradford, 1957
27 Blanc 1953 bis; Bradford, 1957

28 Guy, 1964, 119–21
29 Guy, 1954–5 and 1955
30 Guy, 1964, 120–2
31 In Edelman and Eeuwens (1959), reproduced in Müller-Wille (1970), 29
32 Klinkenberg, 1936
33 Hinz, 1969, Abb. 13–14
34 Schulten, 1898, 12
35 *Corp Inscr Lat* XIII 6488
36 Déléage, 1934, 170
37 Bradford, 1957
38 Suič, 1955
39 Chevallier, 1958
40 1954
41 *Corp Inscr Lat* I^2 585
42 *Corp Inscr Lat* VIII 22786 a–m, 22789
43 *Ann.* 3.73
44 Chevallier, 1958, 101–3
45 W. E. Heitland, 'A great Agricultural Emigration . . .', *J Rom Stud* 8 (1918), 34–52
46 Saumagne, 1929 and 1952; *Atlas des centuriations*, 1954
47 Leschi, 1957
48 Levick, 1967

Chapter 11 The Orange cadasters Pages 159–77

1 Déléage, 1934, pp 73–4, n1
2 We need not follow Oliver (1966) in ascribing the foundation or the intention of it to Julius Caesar
3 Piganiol, 1962
4 2.41.2
5 R. M. Ogilvie, *A Commentary on Livy, Books 1–5* (Oxford, 1965), p 340
6 Singer, 1954, i.792
7 Oliver's article (1966) does not make it clear that Piganiol unequivocally places north at the top of the diagram in Cadaster A
8 Piganiol No 357

Chapter 12 Colonies and State domains Pages 178–87

1 Salmon, 1969, p 15
2 Dittenberger, $Syll^3$ 543
3 See O. A. W. Dilke, 'Do line totals in the *Aeneid* show a preoccupation with significant numbers?' in *Class Quart* 17 (1967), 322–6
4 For the legal status of the settlers, see Rudorff in Blume (1852), 395ff
5 Salmon, 1969

6 Tenney Frank (1919) 205f thought that Piacenza was colonised only in 190 BC and that this earlier foundation was at La Stradella, about 32km west
7 Toynbee, 1965
8 Salmon, 1969, pp 112, 188; in his article in *Athenaeum* 41 (1963), 10f. E. T. Salmon argued that the foundation of Auximum should be dated to 128, not 157 BC
9 Albertini, 1937, 1–4. Appian says these were town boundaries, Orosius and Julius Obsequens, who reflect Livy, say they were field boundaries
10 See bibliography, Chapter 11, North Africa
11 *Corp Inscr Lat* I^2 585
12 For some restorations of the inscription see H. B. Mattingly, 'The two Republican Laws of the *Tabula Bembina*', *J Rom Stud* 59 (1969), 129–43
13 *Annals* 1.17
14 Not necessarily, as thought by Mommsen in Blume ii (1852) 160, a mistake for Julius Caesar: Claudius was a keen writer on antiquarian subjects, and perhaps early agrimensorial lore appealed to him
15 In Blume ii (1852) 158
16 Blume i 230–1
17 Pais, 1923

Chapter 13 Accurate planning in Roman Britain Pages 188–200

1 *Corp Inscr Lat* VII 215
2 *Archaeologia* 42 (1867)
3 *Roman Lancashire* (1883), 233ff
4 Haverfield, 1918
5 Sharpe, 1908, 1918
6 Sharpe, 1932, 83
7 Blume i.110, 6–7
8 Coles, 1939
9 Blume i.158, 20–21
10 Sharpe, 1918, 489–90
11 Nightingale, 1952
12 Margary, 1967, 51
13 Nightingale, 1952, 155 n1
14 Margary, 1940; 1967, 73
15 Coles, 1939
16 Berry, 1949
17 Margary, 1967, 134
18 Richmond, 1945
19 Richmond, 1946
20 Wacher, 1966

21 Margary, 1967
22 Salway *et al*, 1970
23 Stevens, 1966
24 Macdonald, 1934; Robertson, 1963

Chapter 14 Roman surveying and today Pages 201–11

1 Thompson, 1952
2 Beloch, 1890, 66
3 Schmiedt, 1962
4 Sherman, 1925, 221
5 Pattison, 1957, 38 n3
6 Pattison, 1957, 63 n2
7 Levi, 1967
8 Blume ii (1852)
9 Thulin, 1913
10 Castagnoli, 1958
11 Piganiol, 1962
12 Taylor, 1927 and 1956
13 Salmon, 1969

Appendix A

Contents of the Corpus

The authors and treatises comprising the Corpus Agrimensorum, in the order of Blume *et al* i (1848), are as follows:

1. Frontinus, *De agrorum qualitate, De controversiis, De limitibus,* De controversiis agrorum.*† The author is Sextus Julius Frontinus (ca AD 30–104), governor of Britain ca AD 74–8, from whose pen we also possess a work on strategy and a two-book manual on the water supply of Rome. The works in the Corpus which bear his name are incomplete but illustrated. They are on types of land, centuriation, and land disputes, the earliest accounts of these subjects, written by a man of wide experience.

2. Agennius (rather than Aggenius) Urbicus, a late commentary on No 1, *De agrorum qualitate* and *De controversiis.*‡ Illustrated in the Liber Diazographus of the Corpus.

3. Balbus, *Expositio et ratio omnium formarum*, addressed to Celsus. The author's military experience is mentioned on p 42. The addressee may be Publius Juventius Celsus, a well-known legal writer, who was praetor AD 106 or 107 and twice consul, the second time in 129. The word *formarum* is incorrect as applied to this treatise: it should be something like *mensurarum*, since the subject is measurements. Geometrical drawings.

4. Hyginus, *De limitibus, De condicionibus agrorum, De generibus controversiarum.* These and No 6 purport to be written by Augustus' freedman Gaius Julius Hyginus, librarian of the Palatine library. In

* Divided by Thulin into *De limitibus* and *De arte mensoria.*
† Attributed by Thulin to Agennius Urbicus.
‡ Considered by Thulin not to be by Agennius Urbicus.

reality neither can be; the author of these treatises wrote ca AD 100. They concern centuriation, types of holding, and land disputes. The first two are incomplete; there are no illustrations.

5. Siculus Flaccus, *De condicionibus agrorum*. On types of land-tenure in Italy; date uncertain; no illustrations.

6. Hyginus, *De limitibus constituendis*. Usually referred to, on MS authority, as Hyginus Gromaticus,* to differentiate him from the author of No 4. An informative manual on centuriation, written with some pretence at elegance. Date perhaps somewhat later than No 4. Illustrated; one of the more lucid accounts in the Corpus.

7. *Libri coloniarum* I, II. Two lists of colonies with some particulars. The first lists colonies in Italy and Dalmatia, with some technical terms; the second lists colonies in parts of Italy. The first has some illustrations.

8. Three chapters from the Lex Mamilia; no illustrations.

9. Legal extracts from the Codex Theodosianus and other sources; no illustrations.

10. *De sepulchris*. Extracts on burial places; no illustrations.

11. *Agrorum quae sit inspectio*. A short extract on types of land; no illustrations.

12. Marcus Junius Nipsus, *Fluminis varatio*, *Limitis repositio*, *Podismus*. Incomplete; geometrical diagrams; treatise of uncertain date on the geometry of surveying, including estimating the width of a river, and on centuriation.

13. Extracts from Dolabella on problems of surveying; illustrations.

14. Latinus, *De terminibus* (= *terminis*). Two incomplete extracts on boundary stones; illustrated.

15. Gaius; Vitalis; Faustus and Valerius. Short extracts on boundary stones; the first and third are illustrated.

16. *Casae litterarum*. Four lists of country estates under the late Empire, expressed in terms of letters of the alphabet. Three of these are given in Blume *et al* i (1848); Josephson (1950) edits, as the most important, the third of these (III) and another (A). List I is said to come from Innocentius, *De litteris et notis iuris exponendis*; but the Latinity is very late, so that this is not likely to be the surveyor Innocentius mentioned by Ammianus Marcellinus. The MS title of List III is *De*

* So Thulin, who calls the work *Constitutio limitum*.

casis litterarum montium, but it deals with estates on plains too. The word *casa* has developed from classical Latin 'cottage' to 'country house' (hence 'house' in Italian and Spanish), for which the classical Latin was *villa*.

17. *Mensurarum genera*. A table of measurements.

18. *Litterae singulares*. Unexplained letters on certain Italian boundary stones (p 104), mostly two to each stone.

19. *Terminorum diagrammata*. Drawings of boundary stones.

20. *Ordines finitionum*. Short notes on boundary stones; illustrated.

21. Vitalis and Arcadius (cf 26, 25). On boundary stones; illustrated.

22. Gaius and Theodosius. On boundary stones; no illustrations. As with nos 15, 25, 26, the name of a legal writer (Gaius) or an emperor (Theodosius) is given where an extract is made from some legal digest.

23. Latinus and Mysrontius. On boundary stones, monuments etc; no illustrations.

24. *Mago and Vegoia*. Two extracts: (a) on boundaries, aqueducts etc; two illustrations; (b) on religious sanctions against violating boundaries. The former is supposedly from the agricultural work of the Carthaginian Mago. For the date of the latter see p 33 and Heurgon (1959), Pfiffig (1961).

25. The emperor Arcadius (ie his chief surveyor?). A short extract, ca AD 400, on the boundaries of Constantinople; two illustrations.

26. Vitalis; Faustus and Valerius. Short extracts on boundary stones, different from no 15; no illustrations.

27. *Litterae singulares*. Similar to no 18, but a different set; no illustrations.

28. *De iugeribus metiundis*. On measuring the number of *iugera* in fields of various shapes; geometrical drawings.

29. *Litterae singulares*. Single-letter abbreviations, with glosses; no illustrations.

30. *Ratio limitum regendorum*. On boundaries, boundary stones, valleys and marshes.

31. Isidore, *Origines* 15.15, on lands and measurement.

32. An expanded version of no 31, with expanded versions of Isidore, *Origines* 16.25, on weights and measures.

33. Extracts from Euclid's geometry (Latin).

34. Extracts from Cassiodorus' geometry.

35. Extracts from Boethius' geometry.

Not included by Blume *et al* (1848): a fragment attributed to Epaphroditus and Vitruvius Rufus.

Separately edited: *De munitionibus castrorum*, a treatise on camp-making; beginning of work missing; no illustrations. Edited by C. C. L. Lange (1848), who attributed it to Hyginus Gromaticus, and A. von Domaszewski (1887). In the best MS, the Arcerianus, the title is missing, and the wording at the end, 'liber gromaticus Hygini de divisionibus agrorum explicit', does not tally with the contents.

Appendix B

Kardo *and* decumanus

'KARDO' ('CARDO')

There seems a need to re-examine the original meanings of the words *kardo* and *decumanus* as used in surveying. The noun *cardo* properly means 'hinge', but is used of any pivot, axis or pole, including (a) one of the two pivots on which the universe was supposed to rotate round the earth, (b) one of the four cardinal points. The fact that certain terms of surveying were in common with augurs' terms, and that the Corpus traces both back to Etruscan origins, suggests that *cardo* in the surveying sense may have had its origin in augury. Its old spelling *kardo*, for the most part retained in classical and late Latin in this sense, is proof of its antiquity. One cannot imagine the early application of the word arising out of a multiplicity of hinges or the like. There must have been one to start with, which only in contrast with the others became known as *kardo maximus*.

'DECUMANUS' ('DECIMANUS')

Decumanus, or its later equivalent *decimanus*, is properly speaking an adjective derived from *decimus*, 'tenth'. As a surveying term it is used (with the noun *limes* understood) to indicate a *limes*, ie in effect road, at right angles to the *kardines*. Siculus Flaccus, 153 and 136L, has his own theory on the meaning of this term. According to him, supported by Hyginus, 115L, the early *agri quaestorii* were divided into small squares of 50 *iugera* each, with sides of 10 *actus*; this, he argues, caused the roads to be called *decumani* from the number ten. His theory is accepted by Piganiol (1962, 41), the Oxford Latin Dictionary and others. But (a) the terms *kardo* and *decumanus* probably arose not in the measurement of fields, but either as augurs' terms or to denote the main streets of towns; (b) since *cardo* seems to be originally applicable only to one road, or more correctly one *limes*, which then has to be called the main *kardo* to distinguish it from others, one would expect *decumanus* also to apply in origin to only one; (c) there is little evidence that squares of 50 *iugera* with sides of 10 *actus* were standard in the earliest times of centuriation. They are definitely known only from certain allocations in Italy under the Second Triumvirate. The fact that *centuria* was commonly used for the larger unit, and was tied up with the early term *heredium*, meaning 100 of these smallholdings = 200 *iugera*, rather points to sides of 20 *actus* having been standard from the beginning; (d) *decumanus* in the sense 'huge' is common enough in early Latin. It perhaps originated from the idea that every tenth wave is the biggest. But there may also be a parallel in camp terminology.

The gates of a military camp were usually called *porta principalis dextra* and *sinistra*, *porta praetoria* and, opposite this, *porta decumana*. The question arises what this last term really means. The treatise *De munitionibus castrorum*, 18, says it is derived from the tenth cohort, which bivouacked near it. Certainly it is true that the tenth cohort was the nearest to that gate, but scholars are doubtful whether the gate would have been called after an army unit. Some modern writers think it had the same origin as the DM of a land survey. But (a) a camp was not normally laid out with intersecting roads, (b) there is in a camp no *kardo* and no gate named after one. It seems more likely that the *porta decumana* was in origin the widest gate, from which the troops would make sallies, and that again the basic sense is 'huge'.

It remains only to dismiss one unlikely and one impossible etymology. (a) It is not very likely that, because the crossroads looked like the numeral X, this caused the name *decumanus* to be given first to a crossroads and then to a principal road leading from one. (b) The word *decumanus* cannot possibly come from *duo*, indicating bisection, as some writers in the Corpus claim.

It therefore seems reasonable to suggest that *decumanus*, in the sense 'huge', means 'a wide road', as opposed to the narrow *limites* within a 'century'. The widest of the wide roads would come to be called *decumanus maximus*, so that, although at first sight there appears with this interpretation to be tautology, that is not in fact so.

For a discussion of the meanings and uses of *limes* see W. Gebert, 'Limes: Untersuchungen zur Erklärung des Wortes und zu seiner Anwendung', *Bonner Jahrb* 119 (1910), 158–205.

Bibliography

Chapter 1

Oxf Class Dict, 2nd edn, Oxford, 1970, s.v. Gromatici

Chapter 2

Ball, J. *Egypt in the Classical Geographers*, Cairo, 1942

Borchardt, L. 'Ein altägyptisches astronomisches Instrument', Zeitschr f Ägyptische Sprache . . . 37 (1899), 10–17

Breasted, J. H. *Ancient Records of Egypt*, 5 vols, Chicago, 1906–7

Castagnoli, F. *Ippodamo di Mileto e l'urbanistica a pianta ortogonale*, Rome, 1956

Chiera, E. *They Wrote on Clay*, Chicago, 3rd impression, 1939

Déléage, A. 'Les cadastres antiques jusqu'à Dioclétien', *Etudes de Papyrologie* 2 (1934), 73–228

Gadd, G. J. 'Hammurabi and the End of his Dynasty', in *Camb Anc Hist*, rev ed, Vol II, Ch V (1965)

Gardiner, A. H. 'The Inscriptions of Mes', *Untersuchungen zur Altertumskunde Aegyptens* 4 (1905), 1–51

Guy, M. 'L'apport de la photographie aérienne à l'étude de la colonisation antique de la Narbonnaise', in *Colloque International d'Archéologie Aérienne* (Ecole Pratique des Hautes Etudes), Paris, 1964

Hinkle, W. J. *A New Boundary Stone of Nebuchadnezzar I*, Philadelphia, 1907

King, L. W. *The Letters and Inscriptions of Hammurabi*, 3 vols, London, 1898–1900

Lacau, P. and Chevrier, H. *Une chapelle de Sésostris 1er à Karnak*, Cairo, 1956

Loret, V. 'Les grandes inscriptions de Mes à Saqqarah', *Zeitschr f aegypt Sprache* 49 (1901), 1–10

Lyons, Sir Henry G. 'Ancient Surveying Instruments', *Geog J* 69 (1927), 132–43

Lyons, Sir Henry G. 'Land Surveying in Ancient Times', in *Conference of Empire Survey Officers, 1931*, London, 1932

Martin, R. *L'urbanisme dans la Grèce antique*, Paris, 1956
Needham, J. *Science and Civilisation in China*, 4 vols, Cambridge, 1954–62
Neugebauer, O. *The Exact Sciences in Antiquity*, 2nd ed, Providence, R.I., 1957
Oxyrhyncus Papyri VI, ed B. P. Grenfell and A. S. Hunt, London, 1909
Sarton, G. *A History of Science*, 2 vols, London and Cambridge, Mass, 1953, 1959
Schmiedt, G. 'La prospezione aerea nella ricerca archeologica', Fondazione Lerici conference, Venice, 1962
Schmiedt, G. and Chevallier, R. 'Caulonia e Metaponto', *L'Universo* 39 (1959), 349–70
Schmiedt, G. and Chevallier, R. 'Photographie aérienne et urbanisme antique en Grande Grèce . . .', *Rev Archéologique*, 1960, I, 1–31
Singer, C. *et al. A History of Technology*, Vol 1, Oxford, 1954
Taton, R. *Ancient and Medieval Science*, tr A. J. Pomerans, London, 1963
Tebtunis Papyri Pt I (University of California), ed B. P. Grenfell *et al*, 1902
Thureau-Dangin, F. 'Un cadastre chaldéen', *Rev d'Assyriologie et d'Archéol Or* 4 (1897), 13–20
Westerman, W. L. 'Land Registers of Western Asia under the Seleucids', *Class Philol* 16 (1921), 12–19, 391–2
Wycherley, R. E. *How the Greeks Built Cities*, London, 1949 and repr

Chapter 3

Badian, E. 'From the Gracchi to Sulla (1940–1959)', *Historia* 11 (1962), 197–245
Blume, F., Lachmann, K., and Rudorff, A. *Die Schriften der römischen Feldmesser* (Gromatici veteres), 2 vols: I Text, ed Lachmann; II Erläuterungen. Berlin, 1848, 1852, repr 1962
Caterini, R. de. 'Gromatici veteres: i tecnici erariali dell' antica Roma', *Riv del catasto e dei serv tec erar* 2 (1935), 261–358
Chevallier, R. 'Sur les traces des arpenteurs romains', *Caesarodunum* Suppl 2, Orléans-Tours, 1967

Corpus agrimensorum Romanorum i.1 (no more published), ed C. Thulin. Teubner, Leipzig, 1913

Dilke, O. A. W. 'The Roman Surveyors', *Greece and Rome*, ns 9 (1962), 170–80, Pls I–IV

Enciclopedia Italiana s.v. Agrimensura

Frontinus, *Les aqueducs de la ville de Rome*, tr P. Grimal. Budé, Paris, 1944

Frontinus, *The Stratagems* and *The Aqueducts of Rome*, tr C. E. Bennett. Loeb, London and New York, 1925

Hardy, E. G. *Six Roman Laws*. Oxford, 1911

Hardy, E. G. *Three Spanish Charters*. Oxford, 1912

Heurgon, J. 'The Date of Vegoia's Prophecy', *J Rom Stud* 49 (1959), 41–5

Homo, L. *Rome impériale et l'urbanisme dans l'antiquité* (*L'évolution de l'humanité* I.4.III bis). Paris, 1951

MacKendrick, P. L. 'Cicero, Livy and Roman Colonisation', *Athenaeum* 32 (1954), 201–49

Mansuelli, G. A. *Guida alla città Etrusca e al Museo di Marzabotto*. 1966

Nicolet, C. 'Les *finitores ex equestri loco* de la loi Servilia de 63 av. J.-C.', Latomus 29 (1970), 72–103

Nissen, H. *Das Templum*. Berlin, 1869

Oxé, A. 'Die römische Vermessung steuerpflichtigen Bodens', *Bonner Jahrbücher* 20 (1923), 20–7

Pauly-Wissowa, *RE* s.v. Agrimensores, Gromatici

Pfiffig, A. J. 'Eine etruskische Prophezeiung', *Gymnasium* 68 (1961), 55–64

Ruggiero, E. de. *Dizionario epigrafico* s.v. agrimensor. Rome, repr 1961

Salmon, E. T. (1969) See bibliography Chapter 12

Thulin, C. O. 'Die etruskische Disciplin, III', *Göteborgs Högskolas Årsskrift* 15 (1909), no 1; repr Darmstadt, 1968

Tissot, P. de. *Etude historique et juridique sur la condition des agrimensores*, Paris, 1879

Warmington, E. H. (ed). *Remains of Old Latin*, vol IV (Loeb). London and Cambridge, Massachusetts, 1967

Weber, M. *Die römische Agrargeschichte*, 1891, repr Amsterdam, 1962; Italian tr, *Storia agraria Romana*, tr S. Franchi, Milan, 1967, with introduction by E. Sereni

Chapter 4

Blume *et al* (1848, 1852). See bibliography Chapter 3

Boll, F. and Gundel, W. 'Sternbilder, Sternglaube und Sternsymbolik bei Griechen und Römern', Nachträge (with Vol VI) to W. H. Roscher, *Ausführliches Lexikon d gr u röm Mythologie*, cols 869–1071. Leipzig and Berlin, 1924–37

Brugi, B. *Le dottrine giuridiche degli agrimensori romani*. . . . Verona and Padua, 1897

Cantor, M. *Die römischen Agrimensoren und ihre Stellung in der Geschichte der Feldmesskunst*. Leipzig, 1875, repr Wiesbaden, 1968

Caterini, R. de (1935). See bibliography Chapter 3

Columella, *On Agriculture* II (Loeb), tr E. S. Forster and E. H. Heffner. London and Cambridge, Mass, 1954

Dilke, O. A. W. 'Illustrations from Roman Surveyors' Manuals', *Imago Mundi* 21 (1967), 9–29

Hyginus Gromaticus, *Liber de munitionibus castrorum*: (a) ed C. C. L. Lange, Göttingen, 1848; (b) ed G. Gemoll. Teubner, Leipzig, 1879; (c) ed A. von Domaszewski. Leipzig, 1887

Mollweide, C. B. 'Erläuterung einer in der Scriptoribus rei agrariae . . . gegebenen Vorschrift . . .', *Zach's monatliche Correspondenz zum Beförd d Erd- u Himmelkunde* 28 (1813), 396–425

Neugebauer, O. (1957). See bibliography Chapter 2

Nissen, H. *Orientation* I. Berlin, 1906

Rose, H. J. (1923). See bibliography Chapter 6

Tannery, P. *Recherches sur l'histoire de l'astronomie ancienne*. Paris, 1893

Tissot, P. de. *Etude hist et jurid sur la condition des agrimensores dans l'ancienne Rome*. Paris, 1879

Waerden, B. L. van der. *Die Astronomie der Pythagoreer* (Verhandelingen der Kon Ned Akad van Wetenschappen, afd Natuurkunde, 1 xx no 1). Amsterdam, 1951

Weber, M. (1891). See bibliography Chapter 3

White, K. D. *Roman Farming*. London, 1971

Chapter 5

Corte, M. Della. 'Groma', *Monumenti Antichi* 28 (1922), 5–100, reviewed by F. W. Kelsey, *Class Philol* 21 (1926), 259–62

Drachmann, A. G. 'Hero's Instruments', in Singer, C. *et al* (ed), *History of Technology* 3, 609–11. Oxford, 1957

Diels, H. *Antike Technik*. Leipzig and Berlin, 1920
Hero of Alexandria, III. *Rationes dimetiendi et commentatio dioptrica*, ed H. Schoene (*Dioptra*, Greek and German parallel texts, III, 187–315). Teubner, Leipzig, 1903
Kenner, P. Sonnenuhren aus Aquileia. Vienna, 1880
Lancaster Jones, E. in Lyons, Sir H., 'Ancient Surveying Instruments . . .', *Geog J* 69 (1927), 138–41 (on Hero's dioptra)
Legnazzi, E. N. (1887). See bibliography Chapter 6
Lyons, Sir Henry G. (1927). See bibliography Chapter 2
Needham, J. (1954–62). See bibliography Chapter 2
Nowotny, E. *Römische Forschung in Oesterreich, 1912–1924*. Archäol Inst, Vienna, 1925
Price, D. J. de S. 'Portable Sundials in Antiquity, including an Account of a new Example from Aphrodisias', *Centaurus* 14 (1969), 242–66
Schöne, H. 'Das Visirinstrument der römischen Feldmesser', *Jahrb des kais deutsch archäol Inst* 16 (1901), 127–32, Pl II
Shorey, E. N. 'Roman Surveying Instruments', *Univ of Washington Publ in Lang and Lit* 4 (1926–8), 215–42
Singer, C. 'Science', in *The Legacy of Rome*, ed C. Bailey, 265–324. Oxford, 1923
Stebbins, F. A. 'A Roman Sun-Dial', *J of the Royal Astronomical Soc of Canada* 52 (1958), 250–4
Vitruvius, *On Architecture*, Loeb transl Vol II, tr F. Granger. Cambridge, Mass and London, 1934 and repr
Walters, R. C. S. 'Greek and Roman Engineering Instruments', *Trans of the Newcomen Soc* 2 (1921–2), 1–16

Chapter 6

Blume *et al.* See bibliography Chapter 3
Bradford, J. *Ancient Landscapes*, 145–216. London, 1957
Castagnoli, F. *Le ricerche sui resti della centuriazione* (Note e discussioni erudite 7). Rome, 1958
Castagnoli, F. (1956). See bibliography Chapter 2
Castagnoli, F. 'Limitatio', in E. de Ruggiero, *Dizionario Epigrafico s.v. Limitatio* (1964)
Chevallier, R. See bibliography Chapter 3
Déléage, A. See bibliography Chapter 2

Dilke, O. A. W. See bibliography Chapter 4
Enciclopedia Italiana s.v. Agrimensura
Fletcher, E. N. R. 'An Exercise in Ancient Roman Metrology', *Geog J* 133 (1967), 283–4
Grenier, A. *Manuel d'archéologie gallo-romaine* I, 162–85, on boundary stones etc (Paris, 1931); II, 12–23, on roads and survey (Paris, 1934)
Hultsch, F. O. *Griechische und römische Metrologie*, 2nd ed, Berlin, 1882
Jones, H. Stuart. *A Companion to Roman History*. Oxford, 1912
Legnazzi, E. N. *Del catasto romano e di alcuni strumenti antichi di geodesia*. Verona and Padua, 1887
Pauly-Wissowa, *RE* s.v. centuria, Kataster, limitatio
Piganiol, A. (1962). See bibliography Chapter 11
Rose, H. J. 'The Inauguration of Numa', *J Rom Stud* 13 (1923), 82–90
Rossi, G. *Groma e squadro, ovvero storia della agrimensura italiana dai tempi antichi al secolo XVIII*. Rome, Turin and Florence, 1877
Ruggiero, E. De. *Dizionario epigrafico* s.v. adsignatio, centuriatio. Rome, repr 1961. See also Castagnoli (1964)
Salmon, E. T. (1969). See bibliography Chapter 12
Saumagne, C. 'Iter populo debetur', *Rev de Philol*, 3rd ser 2 (1928), 320–58. For Saumagne (1952) see on Chapter 10, North Africa
Schulten, A. 'Die römische Flurteilung und ihre Reste', *Abh Gesellsch Göttingen* n.s. 2, 7 (1898)
Schulten, A. 'Vom antiken Kataster', *Hermes* 41 (1906), 1–44
Thomsen, R. 'The *iter* Statements of the Liber Coloniarum', *Classica et Mediaevalia* 9 (1947), 37–81
Thulin, C. O. (1913). See bibliography Chapter 3, *Corpus*
Warmington, E. H. (ed, 1967). See bibliography Chapter 3

Chapter 7

Alpago-Novello, Luisa. 'Resti di centuriazione romana nella Val Belluna', *Rendic Accad Lincei* 8th ser 12 (1957), 249 ff
Arangio-Ruiz, V. and Pugliese-Carratelli, G. 'Tabulae Herculanenses V' (on boundary disputes), *La Parola del Passato* 10 (1955), 448–77, esp 453–6
Blume *et al*. See bibliography Chapter 3
Brugi, B. 'Fiumi compresi nei lotti dei coloni romani', in *Studi in onore di P. Bonfante* I, 363–6. Milan, 1930
Brugi, B. (1897). See bibliography Chapter 4

Castagnoli, F. 'Cippo di "restitutio agrorum" presso Canne', *Riv Fil Class* 76 (1948), 280–6
Dilke, O. A. W. See bibliography Chapter 4
Mommsen, T. 'Zum römischen Bodenrecht', *Hermes* 27 (1892), 79–117
Pauly-Wissowa, *RE* s.v. Terminatio, Terminus, Terminus motus
Rudolph, H. *Stadt und Staat im römischen Italien*. Göttingen, 1935, repr 1965
Ruggiero, E. de. *Dizionario epigrafico* s.v. cippi, limitatio
Tablettes Albertini: actes privés de l'époque vandale, ed C. Courtois *et al*, Paris, 1952
Thomsen, R. *The Italic Regions* (*Classica et Mediaevalia* diss IV). Copenhagen, 1947
Väänänen, V. *Etudes sur le texte et la langue des Tablettes Albertini* (Suomalaisen Tiedeakatemian Toimituksia B 141, 2). Helsinki, 1965
Warmington, E. H. (ed, 1967). See bibliography Chapter 3

Chapter 8

Bagrow, L. *History of Cartography*, rev R. A. Skelton. London, 1964
Carettoni, G. *et al*. *S.P.Q.R.: la pianta marmorea di Roma antica*. Rome, 1960
Castagnoli, F. 'Le *formae* delle colonie romane e le miniature dei codici dei gromatici', *Atti della Reale Accad d'Italia* (*Mem Linc*, 7th ser) 4 (1943), 83–118
Dilke, O. A. W. 'Maps in the Treatises of Roman Land Surveyors', *Geog J* 127 (1961), 417–26
Dilke, O. A. W. (1962). See bibliography Chapter 3
Dilke, O. A. W. (1967). See bibliography Chapter 4
Dilke, O. A. W. and Margaret S. 'Terracina and the Pomptine Marshes', *Greece & Rome* n.s. 8 (1961), 172–8, Pl V
Grosjean, G., 'La Limitation romaine . . .', *Le Globe*, 95 (1956), 57–74
Heyden, A. A. M. van der, and Scullard, H. H. *Atlas of the Classical World*. London and Edinburgh, 1959 and repr
Levi, Annalina and M. *Itineraria Picta: contributo allo studio della Tabula Peutingeriana*. Rome, 1967
Miller, K. *Die Peutingersche Tafel*, repr Stuttgart, 1962
Notitia Dignitatum, ed O. Seeck, 1876, repr Frankfurt am Main, 1962
Pauly-Wissowa, *RE* s.v. Karten

Richmond, I. A. and Crawford, O. G. S. *The British Section of the Ravenna Cosmography*. Oxford, 1949
Salmon, E. T. See bibliography Chapter 12
Schulten, A. 'Römische Flurkarten', *Hermes* 33 (1898), 534–65
Thomson, J. O. *Everyman's Classical Atlas*. London, 1961
Thomson, J. O. *History of Ancient Geography*. Cambridge, 1948

Chapter 9

Beeson, C. H. 'The Archetype of the Roman Agrimensores', *Clas Philol* 23 (1928), 1–14
Blume *et al* (1848, 1852). See bibliography Chapter 3
Bubnov, N. *Gerberti, postea Silvestri II papae, opera mathematica, 972–1003*. Berlin, 1899
Byvanck, A. W. 'Een antieke miniatur . . .', *Mededeelingen van het Ned Hist Inst te Rome* 3 (1923), 123–36
Casae Litterarum, ed Å. Josephson. Uppsala, 1950
Codices Graeci et Latini photographice depicti: XXII Codex Arcerianus A, ed H. Butzmann. Leyden, 1970
Columella (1954). See bibliography Chapter 4
Corpus Agrimensorum, ed Thulin (1913). See bibliography Chapter 3
Dilke, O. A. W. (1967). See bibliography Chapter 4
Hyginus (1848, 1879, 1887). See bibliography Chapter 4
Josephson, Å. *Casae Litterarum: Studien zum Corpus Agrimensorum Romanorum*. Uppsala, 1950
Lowe, E. A. *Codices Latini Antiquiores* IX. Oxford, 1959
Pais, E. (1923). See bibliography Chapter 12
Thulin, C. 'Die Handschriften des Corpus agrimensorum Romanorum', *Abh preuss Akad Wiss*, phil-hist Cl, Anhang II, 1911

Chapter 10 Part I

Barthel, W. 'Römische Limitation in der Provinz Africa', *Bonner Jahrbücher* 120 (1911), 39–126, Pls I–VII
Bradford, J. (1957). See bibliography Chapter 6
Campana, A. 'Decimo, Decimano, Dismano', in *Emilia Romana* 1. 1 ff. Florence, 1941
Castagnoli, F. (1958). See bibliography Chapter 6

Chevallier, R. *Bibliographie des applications archéologiques de la photographie aérienne*. Fondazione Lerici, Milan, 1957
Chevallier, R. *L'avion à la découverte du passé*. Paris, 1964
Falbe, C. T. *Recherches sur l'emplacement de Carthage*. Paris, 1833
Fraccaro, P. *Opuscula* iii. Pavia, 1957
Hawkes, C. F. C. 'Britons, Romans and Saxons round Salisbury and in Cranborne Chase', *Archaeol J* 104 (1947), 27–81
Kandler, P. *Indicazioni per riconoscere le cose storiche del Litorale*. Trieste, 1855
Legnazzi, E. N. (1887). See bibliography Chapter 6
Mancini, F. *et al. Imola nell' antichità*. Rome, 1957
Nègre, E. *Les noms de lieux en France*. Paris, 1963
Pott, A. F. 'Das Latein im Uebergange zum Romanischen', *Zeitschrift f d Alterthumswissenschaft*, 1854, 219 ff
Richmond, Sir Ian. *Roman Archaeology and Art*. London, 1969
Schmiedt, G. 'Metodi dell' impiego . . . della fotografia aerea . . .', *Atti del settimo Congresso internaz di arch cl*, I, 9–39, Pls I–VIII. Rome, 1961
Schmiedt, G. (1962). See bibliography Chapter 2
Schulten, A. (1898). See bibliography Chapter 6
Seel, K. A. 'Römerzeitliche Fluren im Mayener Stadtwald', *Bonner Jahrbücher* 163 (1963), 317–41
Stevens, C. E. 'Un établissement celtique . . .', *Rev Archéol*, 6th series, 9 (1937), 26–37
Tanioka, T. 'Systèmes agraires: le Jóri dans le Japon ancien', *Annales: Economies* . . . 4 (1959), 625–39
Thulin, C. O. (1909). See bibliography Chapter 3

Chapter 10 Part II

ITALY

There are so many articles on local centuriation in Italy that only a limited selection can be quoted. Most of those listed below are such as are not mentioned by Castagnoli (1958).

Alfieri, N. *et al.* 'Ricerche . . . sul territorio di Loreto', *Studia Picena* 33–4 (1965–6), 1 ff
Bosio, L. 'La centuriazione dell' agro di Iulia Concordia', *Atti dell' Ist Veneto* . . . 124 (1965–6), cl sci mor, 195–260

Bradford, J. 'Buried Landscapes in Southern Italy', *Antiquity* 23 (1949), 58–72

Bradford, J. (1957). See bibliography Chapter 6

Castagnoli, F. 'I più antichi esempi conservati di divisioni . . .', *Bull del Com Prep Ed* 75 (1953–5), Appendix, 3 ff

Castagnoli, F. 'La centuriazione di Cosa', *Mem Amer Acad* 24 (1956), 147–65

Castagnoli, F. *Le ricerche sui resti della centuriazione* (Note e discussioni erudite 7). Rome, 1958

Chevallier, R. 'La centuriazione e la colonizzazione . . . dell' ottava regione Emilia-Romagna', *L'Universo* 40 (1960), 1077 ff; cf *Studi Romagnoli* 13 (1962), 57–83, and Chevallier (1967), bibliography Chapter 3

Chevallier, R. 'La centuriazione . . . dell' Istria e della Dalmazia', *Atti e Mem Soc Istr Arch* 9 (1961), 11–24

Chevallier, R. 'Catasti romani dell' Istria e della Dalmazia', *Boll di Geodesia* 16 (1957), 167 ff

Coarelli, F., in Castagnoli, F. (ed), *Saggi di fotointerpretazione . . .* (Quaderni dell' Ist di Topogr, Università di Roma). 1964

Degrassi, A. *Il confine nord-orientale dell' Italia romana.* Berne, 1954

Fraccaro, P. *Catalogo della Mostra Augustea della Romanità*, 4th ed, 870 ff

Fraccaro, P. *Opuscula* iii. Pavia, 1957

Jones, G. D. B. Forthcoming work on Roman Apulia

Mertens, J. *Alba Fucens I . . .* (Etudes de phil, d'arch et d'hist anc 12). Brussels, 1969

Mertens, J. 'Le système urbain d'Alba Fucens . . .', *L'Antiquité Classique* 27 (1958), 363–72

Pilla, F. G. 'Nota . . . sul rilievamento della centuriazione Trevigiana', *Atti dell' Ist Veneto . . .* 124 (1965–6), cl sci mor, 405–10

Roberti, M. M. *Storia di Brescia* I, Part IV: *Archeologia ed Arte di Brescia Romana*

Salmon, E. T. (1969). See bibliography Chapter 12

Sartori, A. T. 'I confini del territorio di Comum . . .', *Atti del Centro Studi e Documentazione sull' Italia Romana* 1 (1967–8), 275–90

Staedler, E. 'Zu den 29 neu aufgefundenen Inscriftstelen von Minturno', *Hermes* 77 (1942), 149–96

Suič, M. (1955). See bibliography under *Dalmatia*

Tabula Imperii Romani (International Map of the Roman Empire): Mediolanum (1966), Tergeste (1961)

FRANCE

Blanc, A. *Valence romaine*, 18 ff. Bordighera, 1953

Blanc, A. 'Les traces de centuriation romaine et les origines de la cité de Valence', *Rivista Studi Liguri* 19 (1953), 35 ff

Bradford (1957), 207 ff. See bibliography Chapter 6

Braun, J. 'Recherches sur la centuriation romaine dans la plaine d'Alsace à l'ouest de Strasbourg', *Revue archéol de l'Est* 12 (1961), 51

Grenier, A. 'La centuriation romaine de la colonie de Valence', *Gallia* 16 (1958), 281

Guy, M. 'Traces du cadastre romain de quelques colonies de la Narbonnaise', *Etudes Roussillonnaises* 4 (1954–5), 217 ff

Guy, M. 'Vues aériennes montrant la centuriation de la colonie de Narbonne', *Gallia* 13 (1955), 103 ff

Guy, M. (1964) See bibliography Chapter 2

Lebel, P. 'Bornes, centuriation et cantonnement le long de la voie de Lyon au Rhin', *Revue archéol de l'Est* 1 (1950), 154 ff

Lebel, P. 'Les routes. Examen d'un tronçon de voie', *Ann de Bourgogne* 22 (1950), 287 ff

Meynier, A. 'Champs et chemins en Bretagne', *Assoc G Budé*, conf univ de Bretagne, 1943, 161 ff

Meynier, A. 'Traces de cadastre romain en Armorique?' *Comptes Rendus Acad Inscr*, 1944, 413 ff

Musset, L. 'Arpentage antique en Normandie', *Revue Archéol* 28 (1947), 31 ff

Piganiol (1962). See bibliography Chapter 11

THE LOW COUNTRIES

Edelman, C. H. and Eeuwens, B. E. P. 'Sporen van een Romeinse landindeling in Zuid-Limburg', *Berichten van de rijksdienst voor het oudheidkundig bodemonderzoek* 9 (1959), 49–56

Huygen, C. A. 'Romeinse landindeling en Romeinse wegen in Zuid Limburg', *Der Bronk* 7 (1959–60), 353–5

Kakebeeke, A. D. 'Sporen van Romeinse landmeting te Waalre', *Brabants Heem* 20 (1968), 29–37

Mertens, J. 'Sporen van een Romeins kadaster in Limburg?' *Limburg* 37 (1958), 253–60

Mertens, J. 'Enkele beschouwingen over Limburg in de Romeinse tijd', *Archaeologica Belgica* 75 (1964), 5–42

Müller-Wille, M. 'Die landwirtschaftliche Grundlage der Villae rusticae', *Gymnasium* Beihefte, 7.III (1970), 26–42

Ulrix, F. 'Centuriatio in de omstreken van Tongeren', *Limburg* 38 (1959), 34–40

GERMANY

Curschmann, J. 'Die älteste Besiedlung d. Gemarkung Däutenheim bei Alzey', *Mainzer Zeitschr* 18–19 (1921–4), 79 ff

Dopsch, A. *Grundlagen d. europäischen Kulturentwicklung* i, 339 ff. Vienna, 1923

Hinz, H. *Kreis Bergheim* (Archäologische Funde u. Denkmäler des Rheinlandes 2), 60–5. Bonn, 1969

Klinkenberg, J. 'Die Stadtanlage d. röm. Köln u. die Limitation d. Ubierlandes', *Bonner Jahrb* 140–1 (1936), 259–98

Müller, R. 'Die Geographie d. Peutingerschen Tafel in d. Rheinprovinzen, in Holland u. Belgien', *Geogr Anz*, 1926, Heft 9/10

Schmitz, H. *Stadt u. Imperium: Köln in röm Zeit*, 51 ff. Cologne, 1948

Schmitz, H. *Colonia Claudia Ara Agrippinensium*. Cologne, 1956

Schulten, A. 'Flurteilung u. Territorien in d. röm. Rheinland', *Bonner Jahrb* 103 (1898), 12–41

Schumacher, K. 'Beiträge zur Siedlungs- u. Kulturgeschichte Rheinhessens', *Mainzer Zeitschr* 15–16 (1920–1), 1 ff

Schumacher, K. *Siedelungs- u. Kulturgeschichte d. Rheinlande* ii.221–3. Mainz, 1923

SWITZERLAND

There is quite an extensive bibliography, which is given in Grosjean (1963); earlier items are therefore omitted.

Déléage, A. (1934). See bibliography Chapter 2

Grosjean, G. 'Die römische Limitation um Aventicum und das Problem der römischen Limitation in der Schweiz', *Jahrb d Schweizerische Gesellsch f Urgeschichte* 50 (1963), 7–25

Laur-Belart, R. 'Eine römische Landkarte von Aventicum', *Geneva* n.s. 11 (1963), 95–104

DALMATIA

Bradford, J. 'A Technique for the Study of Centuriation', *Antiquity* 21 (1947), 197 ff

Bradford, J. (1957), 178 ff. See bibliography Chapter 6

Chevallier, R. (1957, 1961). See bibliography under *Italy*

Suič, M. 'Limitation of Roman Colonies on the Eastern Adriatic Coast', *Zbornik instituta za historijske nauke u Zadru*. Zadar, 1955

GREECE

Chevallier, R. 'Pour une interprétation archéologique de la couverture aérienne grecque . . .', *Bull Corr Hell* 82 (1958), 635–6

NORTH AFRICA

Atlas des centuriations romaines de Tunisie, edited for the Institut Géographique National. Paris, 1954

Barthel, W. 'Römische Limitation in der Provinz Africa', *Bonner Jahrb* 120 (1911), 39 ff

Bradford, J. (1957), 193 ff. See bibliography Chapter 6

Caillemer, A. and Chevallier, R. 'Les centuriations de l'Africa Vetus', *Annales: Economies* . . . 9 (1954), 433 ff

Caillemer, A. and Chevallier, R. 'Les centuriations romaines de Tunisie', *Annales: Economies* . . . 12 (1957), 275 ff

Caillemer, A. and Chevallier, R. 'Die römische Limitation in Tunesien', *Germania* 35 (1957), 45–54

Chevallier, R. 'Fotografia aerea e archeologia: applicazione allo studio dei catasti antichi', V° Congresso della SIFET. Palermo, 1957

Chevallier, R. 'Essai de chronologie des centuriations romaines de Tunisie', *Mélanges Ec Franç Rome* 70 (1958), 61–128

Davin, P. L. 'Note sur le cadastre romain du Sud Tunisien', *Bull Archéol du Comité des Travaux Hist et Scient*, 1930–1, 689 ff

Falbe, C. T. *Recherches sur l'emplacement de Carthage*, 54 ff. Paris, 1833

Horlaville, Chief Engineer. 'Plan d'un lotissement antique dans la presqu'île du Cap Bon . . .', *Bull Archéol du Comité des Travaux Hist et Scient*, 1950, 125 ff

Legendre, M. 'Note sur la cadastration romaine de Tunisie', *Cahiers de Tunisie* 5 (1957), 125

Leschi, L. 'Une assignation de terres en Afrique . . .', *in Etudes d'épigraphie, d'arch et d'hist afr* 75–9. Paris, 1957

Poncet, J. 'Vestiges de cadastration antique et histoire des sols en Tunisie', *Cahiers de Tunisie* 1 (1953), 323 ff

Saumagne, C. 'Les vestiges d'une centuriation romaine à l'est d'El-Djem', *Comptes Rendus Acad Inscr*, 1929, 307–13

Saumagne, C. 'La photographie aérienne au service de l'archéologie en Tunisie', *Comptes Rendus Acad Inscr*, 1952, 287 ff

Schulten, A. (1898), 36 ff. See bibliography Chapter 6

Schulten, A. 'L'arpentage romain en Tunisie', *Bull archéol du Comité des Travaux Hist et Scient*, 1902, 129 ff

Toutain, J. 'Le cadastre de l'Afrique romaine', *Mém Acad Inscr* 12, 1 (1907), 341 ff; cf *Mém Soc Antiq de France* 10 (1912), 79 ff

ASIA MINOR AND ADJACENT AREAS

Levick, Barbara. *Roman Colonies in Southern Asia Minor*. Oxford, 1967

Chapter 11

Déléage, A. (1934). See bibliography Chapter 2

Oliver, J. H. 'North, South, East, West at Arausio and elsewhere', in *Mélanges d'archéologie et d'histoire offerts à A. Piganiol* 2, 1075–9. Paris, 1966

Piganiol, A. *Les documents cadastraux de la colonie romaine d'Orange. Gallia*, Suppl 16 (with full bibliography). Paris, 1962

Richmond, I. A. and Stevens, C. E. 'The Land Registers of Arausio', *J Rom Stud* 32 (1942), 65–77

Singer, C. *et al* (1954). See bibliography Chapter 2

Chapter 12

Afzelius, A. *Die römische Eroberung Italiens*. Copenhagen, 1942

Albertini, E. 'Les loups de Carthage', in *Mélanges de géographie . . . offerts à M. E. F. Gautier*. Tours, 1937

Blume *et al* (1848, 1852). See bibliography Chapter 3

Carcopino, J. *Autour des Gracques*. Paris, 1928

Frank, Tenney. 'Placentia and the Battle of the Trebia', *J Rom Stud* 9 (1919), 202–7

Pais, E. *Storia della colonizzazione di Roma antica* (contains text of *Libri coloniarum*). Rome, 1923

Ruggiero, E. De. *Dizionario epigrafico* s.v. colonia

Salmon, E. T. *Roman Colonization under the Republic*. London, 1969

Toynbee, A. J. *Hannibal's Legacy*, 2 vols. London, 1965

Chapter 13

Berry, C. A. F. 'Centuriation', *Trans Bristol and Gloucestershire Archaeol Soc* 68 (1949), 14–21

Boon, G. C. 'Belgic and Roman Silchester . . .', *Archaeologia* 102 (1969), 1–81

Blume, F. *et al* (1848, 1852). See bibliography Chapter 3

Coles, R. 'Centuriation in Essex', *Essex Naturalist* 26 (1939), 204–20

Curwen, E. C. *The Archaeology of Sussex*, 2nd ed. London, 1954

Haverfield, F. 'Centuriation in Roman Britain', *Engl Hist Rev* 33 (1918), 289–96 = 'Centuriation in Roman Essex', *Trans Essex Archaeol Soc* n.s. 15 (1918–20), 115–25

Macdonald, Sir G. *The Roman Wall in Scotland*, 2nd ed, Chapter X. Oxford, 1934

Margary, I. D. 'Roman Centuriation at Ripe', *Sussex Archaeol Collections* 81 (1940), 31–41

Margary, I. D. *Roman Roads in Britain*. London, rev 1967

Miller, C. 'On Roman Roads in Essex', *Trans Essex Archaeol Soc* n.s. 15 (1919–20), 190–229, esp 225–7

Nightingale, M. D., with introd by Stevens, C. E. 'A Roman Settlement near Rochester', *Archaeologia Cantiana* 65 (1952), 150–9

Richmond, I. A. 'The Sarmatae, Bremetennacum Veteranorum and the Regio Bremetennacensis', *J Rom Stud* 35 (1945), 15–29

Richmond, I. A. 'The four Coloniae of Roman Britain', *Archaeol J* 103 (1946), 57–84

Richmond, I. A. and Crawford, O. G. S. *The British Section of the Ravenna Cosmography*. Oxford, 1949

Rivet, A. L. F. *Town and Country in Roman Britain*, rev ed. London, 1964

Robertson, Anne S. *The Antonine Wall*, 2nd ed. Glasgow, 1963

Salway, P. *et al. The Fenland in Roman Times,* ed C. W. Phillips. R.G.S. Research Series No 5. London, 1970

Sharpe, M. *Roman Centuriation of the Middlesex District.* Brentford, 1908; cf the same author's 'Centuriation in Middlesex', *Engl Hist Rev* 33 (1918), 489–92, and *Middlesex in British, Roman and Saxon Times,* 2nd ed. London, 1932

Stevens, C. E. *The Building of Hadrian's Wall.* Cumberland and Westmorland Antiquarian and Archaeol Soc, Extra Ser 20. Kendal, 1966

Wacher, J. S. (ed). *The Civitas Capitals of Roman Britain.* Leicester, 1966

Chapter 14

Atlas des centuriations romaines de Tunisie (1954). See bibliography Chapter 10 (*North Africa*)

Beloch, J. *Campanien,* 2nd ed. Breslau, 1890

Blume, F. *et al* (1848, 1852). See bibliography Chapter 3

Bradford, J. (1957). See bibliography Chapter 6

Castagnoli, F. (1958). See bibliography Chapter 6

Chevallier, R. (1967). See bibliography Chapter 3

Corte, M. Della (1922). See bibliography Chapter 5

Dilke, O. A. W. (1967). See bibliography Chapter 4

Goesius, W. *Rei agrariae auctores legesque variae. . . .* Amsterdam, 1674

Levi, Annalina and M. (1967). See bibliography Chapter 8

Pattison, W. D. *The Beginnings of the American Rectangular Land Survey System.* Chicago, 1957

Piganiol, A. (1962). See bibliography Chapter 11

Richeson, A. W. *English Land Measuring to 1800: Instruments and Practices.* Cambridge, Mass and London, 1966

Salmon, E. T. (1969). See bibliography Chapter 12

Schmiedt, G. (1962). See bibliography Chapter 2

Schöne, H. (1901). See bibliography Chapter 5

Sherman, C. E. *Ohio Land Subdivisions.* Columbus, Ohio, 1925

Taylor, Eva G. R. 'The Earliest Account of Triangulation', *Scot Geog Mag* 43 (1927), 341–5

Taylor, Eva G. R. *The Haven-Finding Art: a History of Navigation from Odysseus to Captain Cook.* London, 1956

Thompson, E. A. (ed). *A Roman Reformer and Inventor, being a new text of the treatise De rebus bellicis*. Oxford, 1952
Thulin, C. (ed, 1913). See bibliography Chapter 3, *Corpus*
Warmington, E. H. (ed, 1967). See bibliography Chapter 3

Note. The items listed above which have already been mentioned in bibliographies to earlier chapters are among those which form a general introduction to the subject of Roman surveying.

Acknowledgements

I am much indebted to Professor F. Castagnoli, of the University of Rome, for many references and illustrations, especially on the centuriation of Italy; to Professor R. Chevallier, of the University of Tours-Orléans, for help with centuriation in France, Italy and Tunisia; to Dr J. B. Harley, of the University of Exeter, for editorial help; to Dr G. D. B. Jones, of the University of Manchester, for help with the topography and centuriation of Apulia; to General Giulio Schmiedt, of the Istituto Geografico Militare, for references to Greek colonies; to Mr C. E. Stevens, of Magdalen College, Oxford, for help with Roman Britain and for the loan of cartographic material; to Mr J. B. Ward-Perkins, Director of the British School in Rome, for facilities placed at my disposal and for help with points of topography; and to my colleague Professor H. B. Mattingly for reading through the historical sections.

For photographic reproduction and permission to reproduce diagrams I am indebted to the Biblioteca Apostolica Vaticana and to the Herzog August Bibliothek, especially to its former director Dr H. Butzmann. For permission to reprint photographic or other material I wish to acknowledge the help of *Imago Mundi* and its editor Dr C. Koeman; Dr I. D. Margary; Mrs Bleeker-Eeuwens; *Caesarodunum*; *Gallia*; the *Geographical Journal*; *Greece & Rome*; the British School in Rome; the Centre National de la Recherche Scientifique; the Institut Géographique National; the Science Museum, London; the Ordnance Survey; and the Ministry of defence.

For help with surveying in Ptolemaic Egypt I wish to thank Dr P. Parsons, of Christ Church, Oxford; for museum assistance the Museo Nazionale, Naples, the Museo Archeologico, Aquileia, the Science Museum, London, and the Oxford Museum of the History of Science; for references to Roman Britain, my colleague Mr B. R.

Hartley; and for bibliographical references Mr K. D. White, of the University of Reading, and Mr M. R. Colton.

My wife has given encouragement and help of every kind throughout, and my son Stephen W. Dilke worked with me on the mathematics of Columella's farming book. Mrs J. K. Izatt has given secretarial assistance.

Leeds, January 1971

Index

Abbreviations, 90–2
Acilius Glabrio, M', 43
acnua, 33, 83
actus, 70, 82–3, 85, 139, 142, 146–7, 193–5
Admedera, *see* Ammaedara
Ad Tricesimum, 85, 138, 148
Aebutius Faustus, L., 39, 50, 66
Aecae (Troia), 144
Africa, north, 36–7, 40, 45, 65, 85, 93–4, 110, 134, 151–8, 183
Agathe Tyche (Agde), 25
Agennius Urbicus, 105–8, 126, 130
agentes in rebus, 44–5
ager arcifinius, 96, 106
ager publicus, 88, 106, 139, 160, 178, 181
Agrarian laws, 36, 106, 112, 156, 160
Agriculture, 34, 52–5, 157, 176
agrimensores, 15–18, 25, 31ff *passim*; training, 47–65
Agrippa, M. Vipsanius, 109, 111
Aigue, R., 164
Alba Fucens, 94, 136, 146, 180
Albertini tablets, 103
Alexander the Great, 26–7
Alexandria, 26–7
Algebra, 56
Algeria, 103, 157
Allifae (Alife), 88, 144
Allocation of land, 96–7
Altars, 99, 144
Altinum (Altino), 85, 148
Alzey, 150
America, 203–6
Ammaedara (Haïdra), 68, 88, 122, 155, 157
Anticyra, 43
Antium (Anzio), 179
Antonine Itinerary, 122; Wall, 199–200
Antoninus Pius, 51, 199
Anxur, *see* Tarracina
Apollonius of Perga, 57
Apulia, 136, 138, 142, 144, 146
Aquileia, 71–2, 148, 181
Aquinum (Aquino), 85, 135, 144
Arausio, *see* Orange
Arbitrators, *see* Judges
Arcadius, 44
Archimedes, 62, 74
Architects, 48–9
Archytas, 22
areae, 175
Arelate (Arles), 149
Arithmetic, 47
Army, 40, 42, 86; *see also* Camps, Legions
Asculum Picenum, 187
asteriskos, 69–70
Astronomy, 61–3, 75, 78
Athens, 25–6, 108
Atina, 146
Augurs, 32–5, 85, 89
Augusta Emerita (Mérida), 107
Augusta Praetoria (Aosta), 123, 172, 184
Augusta Rauricorum (Augst), 150
Augusta Taurinorum (Turin), 123, 149, 171

Augustodunum (Autun), 109
Augustus, 37–9, 103, 109, 123, 148, 177, 183–4
Aurelius, M., 51; column, 44
Auspices, 32–5, 57
Aventicum (Avenches), 123, 150, 172
Avignon, 149, 176

Babylonia, 19–20, 98
Baeterrae (Béziers), 149
Baiae, 46
Bainbridge, 69
Balbus, surveyor, 38, 42, 60, 130–1, 185
'Begoe' (Vegoia), 33, 126
Belgium, 149–50
Belunum (Bellunum), 85, 148
Beneventum, 85, 107
Berre, R., 168, 176
Bithynia, 42
Bobbio, 128
Boëthius, 127
Bonificazioni, 139
Bononia (Bologna), 135, 148, 180–1
Books, 132; *see also* Corpus, *libri*
Boundaries, 98–108; boundary disputes, *see* Disputes
Boundary stones, 43, 92–3, 98ff, 186
Box, surveyor's, 70–3
Britain, 110–12, 136, 139, 184, 190–200, 210–11
Brixia (Brescia), 149
Building surveyors, 42
Bureaucracy, 45

Caburrum (Cavour), 149
Cadasters, Orange, 41, 84, 159–78, 212; Cadaster A, 163–5; B, 165–73; C, 173
Caesar, Julius, 37, 104–5, 123, 144, 151, 183
Caesarius, Abbot, 46
Cales (Calvi), 94, 144, 180
Campania, 85, 87, 135, 142, 144, 207
Camps, 66, 86, 122, 132–3, 195–6
Canada, 205
Cannae, 108
Capua, 144, 207
cardo, *see kardo*
Carpentras, 173
Carthage, 34, 36–7, 134, 156, 182–3
Cartography, *see* Maps
Casae litterarum, 130
Cassiodorus, 21, 45–6, 108, 128
Cato, elder, 34
Caulonia, 25
centuria, 'century', 15–16, 83ff, 113–14, 133ff
Centuriated areas, 18, 41ff, 143, 145, 147, 152, 155; *see also* Africa, Dalmatia, Gaul, Greece, Italy, Rochester
Centuriation, 82ff, 154, 203, 213; origins and causes, 33–4, 133–4; unorthodox systems, 138–41; work on centuriation, 134–41; *see also* Centuriated areas
Centuriation stones, 89
Chests, 73
Chichester, 198
China, 76
chorobates, 74–6
Christ, Christians, 45–6, 51
Cicero, 37, 183
cippi, *see* Stones
Circeii, 179
Cisalpine Gaul, 147, 181
Clastidium, 148
Claudius, emperor, 183–5
Claudius, Appius, 182
Cliffe, Kent, 191–3
Colchester, 184, 195
collegia, 40
Cologne, 150, 184
Colonia Augusta, 123, *see also* Augusta; Claudia, 123, 150, 184; Flavia, 59; Iulia, 122–3, 159, 183
Colonies, Greek, 24–5, 34, 94, 138, 178; Latin and Roman, 68, 107, 112, 125, 133, 159–87
coloni maritimi, 180
Colours in MSS, 131
Columban, 128
Columella, 33, 52–5, 84, 132
Common land, *see* Pastures

Compasses, 73
Compass points, 103, 163, 166, 173
compitum, 144
Cora (Cori), 178
Corinth, 37, 151, 183
Corpus Agrimensorum, 17, 37–45, 79, 84–6, 90–158 *passim*, 178–87; contents, 227–30
Corsica, 41
Cosa, 94, 136, 146, 180
Cosmology, 61–3
Cremona, 85, 135, 148–9, 180
cultellare, 59
Curtius Rufus, Q., 173
Cyrenaica, 96

Dacia, 42
Dalmatia, 56, 87, 136, 142, 150–1
Danube, 112, 126
decempeda, 67, 73
decempedator, 37
Decidius Saxa, L., 37
decumanus (*decimanus*), 38, 68, 86ff, 124, 137, 146, 160; meaning, 231–2
decuria, 133
Delphi, 43
De iugeribus metiundis, 44
Demonstratio artis geometricae, 37
Dertona (Tortona), 148
De terminibus (*sic*), 103
Diocletian, 44, 151
Diophantus, 56
dioptra, 40, 75–9
Dismano, 137
Disputes, boundary, 44–6, 105–8
Distance calculation, 59–61
Division of land, 87–96
Dolabella, writer in Corpus, 98–9
Domitain, 41–2, 94, 159, 175
Donzère, 166
Dura Europus, 112

Education, *see* Training
Egypt, 19–22, 26–30, 73
Emilia, 146
Enfida, 155
Epagathus, surveyor, 157
Epidaurum, 151
Eporedia (Ivrea), 66, 149
Erasmus, 128
Eratosthenes, 26, 61, 109
Erfurt, 129
Etruria, 146; Etruscans, 32–4, 89
Euclid, 26
evocatus, 42, 157

Falerio, 94, 102
Fanum Fortunae, 146
Farms, 101, 108, 113–14, 125, 138
Faustina, 114
Fayyûm, 27–30; instrument, 27, 49, 69–70
Feltria (Feltre), 148
Fenland, 198–9
Fidenae (Castel Giubileo), 178
Field patterns, 138–41
fines, 137
finitores, 35–6
Florence, 129, 135–6, 146
Foligno, 121–2
Forests, 125
forma, *see* Maps
Forma Urbis Romae, 112–13, 119, 212
Forts, *see* Camps
Forum Domiti(i), 149; Iuli(i) (Cividale), 85, 148; Popili, 147
Foss(e) Way, 197–8
Freedmen, 39–40
Friedberg, 150
Frontiers, 54, 136
Frontinus, Sex. Julius, 40–1, 64–5, 86, 95–6, 105–8, 126, 129–32, 203
Fylde, 195

Gargrave, 141
Gaul, 148–9, 159–77; *see also* Cisalpine, Transpadane
Gellius Sentius Augurinus, Q., 43
Genoa, 100
Geodesy, 26
'Geological map', 19
Geometers, 44, 67

Geometry, 22, 44, 51–61
Gerbert, Abbot, 17, 128
Germany, 42, 54, 139–41, 150
Gloucester, 195
Goes, W., 204
Gracchi, 36–7, 87–8, 92–3, 103, 144, 146, 160, 186; C. Gracchus, 134, 156, 182; Ti., 181–2
Graviscae, 181
Greece, 22–7, 34, 151
Greek colonies, *see* Colonies
Gregory I, Pope, 46
groma, 15–16, 33, 39, 50, 66–70, 89
gromatici, 223; *see also agrimensores*, Corpus
Guilds, *see collegia*

Hadrian, 43, 157, 184–5
Hadrumetum (Sousse), 155
Hammurabi, 19
Hasta (Asti), 171
Heraclea (Gulf of Taranto), 25
heredium, 83, 133
Herodotus, 20
Hero(n), 40, 45, 54, 70, 73, 75–9
Hippo Diarrhytus (Bizerta), 152
Hippodamus, 23, 34
Hispellum (Spello), 118, 120–1, 146, 154
Hodometer, 79–80
Holland, 139–40, 149–50, 203–4
Horace, 47
Hundreds, 204–5
Hyginus, 38, 96–7, 100, 130, 132
Hyginus, C. Julius, 38
Hyginus Gromaticus, 38, 51, 56–9, 62–3, 68, 84, 86, 88, 91–2, 95–7, 99, 113, 115–17, 120, 122, 130–2, 180, 185
Hymettus, Mt, 141
Hypata, 43

Iader (Zadar), 87, 150–1
Ikhnaton, 22
Illustrations, MS, 112–25, 127–32
Instruments, surveying, 66–81, 212–13
Interamna, 180
Interamnium Praetutianum, 107
Irrigation, 141
Isidore, 127
Islands, 173–4
Istria, 148
Italy, 65, 85, 133–9, 183; map, 143; north, 36, 100, 137, 142–9; regions, 99–100, 186
iter, 93
Itineraries, 112, 132
iugerum, 83–5, 96–7, 114–15, 133ff, 169, 179, 191
Iulia Concordia (Concordia), 148
Iulium Carnicum (Zuglio), 148

Jabron, R., 168
Japan, 139
Jefferson, 204–5
Jena, 129–30
John, Bishop of Syracuse, 46
John, surveyor, 46
Jóri, 139
Judges, surveyors as, 44, 103ff
Julius Victor, 43; *see also* Frontinus, Hyginus
Junonia, *see* Carthage
Justinian, 45

kalendarium, 173
kardo, kardines, 38, 68, 86ff, 124, 136–7, 160, 169, 175; meaning, 231
Kerkeosiris, 27–9
Kreuznach, 150

Lakes, 122
Lamia, 43
Land area, 106; classification, 63–5; commissions, 35–6, *and see tresviri*; occupation, ownership, position, 106
Lands, *see ager*, Agrarian laws, Agriculture, Field patterns, Pastures
Lanuvium, 186
latifundia, 181

Latin league, 179
Latinus on boundaries, 184
Latium, 43, 142
Laus Pompeia (Lodi), 149
Law, 44–5, 51, 63–5, 186, 203
Ledgers, 97
Legionaries, ex-, *see evocatus*
Legions: II Augusta, 159, 199; III Augusta, 40, 93, 156; VI, XX, 199
Lettering, 176–7
Levelling, 59, 74–9
Lex Mamilia, 44, 103–6, 126; 'Thoria', *see* Agrarian laws
Lez, R., 168
libri aeris, 113–14
Libri coloniarum, 84, 93, 100, 185
Limburg, 139–40, 149–50
limes, limites, 15, 38, 45, 88ff, 99, 106, 108, 134, 136–7, 139, 176, 191, 233
limitatio, see Centuriation
Lincoln, 195
Liris, R., 153
Liternum, 187
Livius Drusus, M., 182
Livy, 116, 147–8, 160, 180
London, 197
Lottery, 96–7
Luca (Lucca), 88, 146
Lucan, 61–2
Lucania, 146
Luceria (Lucera), 85, 136, 144, 146
Luna (Luni), 146
Lynchets, 139

Maastricht, 150
Macedonia, 43
magistri, 144
Mago, 34, 126
Mainz, 150
Manchester, 190
Mantua, 149, 184
Manuals, surveying, 126–32, 227–30
Manuscripts, 127–31, 212
Maps, mapping, 108–25: Egyptian, 19: Babylonian, 19–20
Marzabotto, 33–4
Massicus, Mt, 122, 171
Mathematics, 51–2; *see also* Algebra, Arithmetic, Geometry
Mauretania, 184
Measurement, 82–5, 162, 203, 210; Egyptian, 21; instruments, 67, 73
Mediolanum (Milan), 149
Mejerda, 155
merides, 175
Mes, tomb of, 22
Mesopotamia, *see* Babylonia
Metapontum, 25
metator, 36–7
Metellus Sequanus, 129
Miletus, 22–4
Military surveyors, *see* Army, Camps, Legions
Minturnae, 117–18, 133, 144, 153
Minucii, Q. and M., 36
Montélimar, 175
Museums, 72, 135–6, 212–13
municipium, 107–8, 125, 178
Mutela, Mt, 125
Mutina (Modena), 147–8, 181

Names, 162–3
Naples, 202, 208; museum, 50, 69–73
Narbo (Narbonne), 149, 182–3
Near East, 19–20; *see also* Babylonia, Syria
Nepet (Nepi), 179
Nero, 40, 79
Nerva, 41
Newton Kyme, 195
Nicaea (Asia Minor), 23
Nicopolis, 151
Nile, 20–2
Nipsus, M. Junius, 55–6, 59–61, 95, 130
Nola, 144
Notitia dignitatum, 110–12, 132
Novaria (Novara), 149
Noviodunum (Nyon), 150
Nuceria (Nocera), 144
Numerals, 34, 179

Octavian, *see* Augustus

Opitergium (Oderzo), 148
Orange, 17, 41, 84, 87, 114–15, 119, 135–6, 142, 149, 159–77, 185, 189–90
Orientation, 56–8, 70, 86–7, 163ff
Ose, R., 118, 121
Ostia, 43, 179
Ouvèze, R., 164
Ovid, 31–2, 105
Oxford sundial, 72
Oxyrhynchus, 29–30

Paestum, 180
Palermo stone, 21
Pannonia, 42–3
Papyrus, 132
Parchment, 132
Parentium (Poreč), 148, 151
Parma, 147–8, 181
passus, 82
Pastures, 107, 113, 124, 172
Patavium (Padua), 94, 148
Pella, 151
Perspective, 124
pertica, 67, 73, 137, 148, 210
pes, 73, 82
Peutinger Table, 109–11, 212
Pfeddersheim, 150
Pfünz instrument, 67, 69–70
Pharos Island (Hvar), 85, 151
Philip V, 178
Photography, air, 136, 141, 211
Picenum, 186–7
Pincers, 73
Pisa, 136, 146
Pisaurum, 181
Pisaurus, R., 101, 107
Place-names, 121, 137–8
Placentia (Piacenza), 33, 146, 148, 180–1
Planning, *see* Town planning; in Britain, 188–200
Plans, *see* Maps
Pliny, elder, 71, 79, 84; younger, 42, 185
Plumb-bobs, plummets, 69–70, 76, 79
Po, 107; valley, 84, 87, 134, 138, 142, 146–8
Poetovio (Ptuj), 43
Pola (Pula), 148, 151
Polentia (Pollenzo), 123, 148
Polybius, 132
Pompeii, 50, 69–73
Pomptine (Pontine) marshes, 115–16, 120
pontifex, 33
Population, 148, 151
possessio, 106
possessores, 181
Potentia, 181
primicerius, 44–5
Ptolemies, 27, 96
Ptolemy, geographer, 17, 109, 111
Pyramids, 21–2
Pythagoras, 22, 55–6, 63
Pythagoreans, 62
Pytheas of Marseilles, 26

Quinta, Quintanilla, 138
quintarii, 68, 93
Quintilian, 47
Quinzano, 138

Rainwater, 65, 108
Ravenna, 137; cosmography, 112
Religious cults, 107–8, 186
Rents, 162, 169
Ribchester, 195
rigor, 106
Rhône, 142, 164–6, 169–70, 173, 176, 190
Ripe, Sussex, 193–5
Rivers, 42, 59–61, 106–7, 125; *see also* Aigue, Ana, Berre, Jabron, Lez, Liris, Nile, Ose, Ouvèze, Pisaurus, Po, Rhône
Roads, 41–2, 108, 122, 127, 132, 148; Via Aemilia, 54, 87, 145–8, 181; Via Appia, 42, 87–8, 115–17, 120, 180; Via Domitiana, 41; Via Postumia, 87; Via Traiana, 146
Rochester, 191–3
Rock cutting, 42
Rome, 42–4, 109, 113, 135–6

Rotteln, 141
Ruler, 73
Rullus, P. Servilius, 35–6

Sabine hills, 125
Salaries, 44
Salonae, 151
saltus, 191
Samnium, 146
Samos, 22
Saturnia, 181
scamna, 63–4, 94–5
Scolacium (Sqillace), 182
Segusio (Susa), 123
Selinus, 25
Septicius, 125, 138
Septimius Severus, 157
Sesostris, 20
sevir, 39
Sextantio, 149
Siculus Flaccus, 44, 63, 65, 98, 130, 150
Sighting, 58–9, 70, 77
Signia (Segni), 178
Silchester, 196–7
Siliana, Wadi, 155
Silvanus, 98–9
Sittard, 150
Spain, 107–8, 110, 138, 142
Spello, *see* Hispellum
Square, carpenter's, 76
Squares, non-Roman, 139, 203–6
State lands, 125, 161–2, 178–87; *see also* *ager publicus*
Statius, 41, 127
Stella, 39, 69
Stones, *see* Boundary, Cannae, Centuriation, Tombstone
strigae, 63–4, 94–5
Strips, 94
subseciva, 94, 99, 101, 107–8, 162
Suessa Aurunca, 95, 122, 171
Suetonius, 183
Sufetula (Sbeitla), 155
Sulla, 183
Sundials, 57, 70–3, 86
Sutrium (Sutri), 179
Syracuse, 46

Tables, Twelve, 106
Tablets, Albertini, 103; bronze, 112–14, wax, 73
tabularium, 113, 160, 173–4
Tacfarinas, 156–7
Tacitus, 184
Tarentum (Taranto), 182
Tarracina, 42, 87–8, 115–17, 120, 142, 179–80
Tarvisium (Treviso), 85, 148
Taxation, Egyptian, 20–1; *see also* Tribute
Tebtunis, 27–30
Technical education, *see* Training
Tegurini, 172
Temples, 86
Tergeste (Trieste), 135, 148
terminatio, 99
Terminus, *see* Boundary stones
Terminus, god, 98
territorium, 100, 107
Thessalonica, 151
Tiberius, 40, 135, 156–7
Ticinum (Pavia), 149
togatus Augusti, 45
Tombstone, 66
Tongeren, 150
Town-planning, town plans, 23, 115, 138, 159–60
Tragurium (Trogir), 151
Training of surveyors, 47–65
Trajan, 41–3, 51, 116, 126, 175, 184; column, 42
Transpadane Gaul, 148
Treatises, *see* Manuals
Trees, 98, 101, 103, 108, 176
tresviri, 35–7, 92, 182
Tribute, 161, 169
Tricastini, 159–77 *passim*
Tricesimo, *see* Ad Tricesimum
Triumvirs, 37, 184
Tunisia, 40, 85, 103, 122, 135–8, 142, 151–8, 190
Tunnels, 22

Turin, *see* Augusta Taurinorum

Umbria, 146
United States, *see* America
Urso (Colonia Iulia), 32, 37, 183
Utina (Udine), 148

Valentia (Valence), 136, 149
Valentinian II, 44
Vandals, 103, 158
Varro, 32, 87
Vatican MSS, 17, 68, 109, 120, 123–4, 129–31, 153–4, 171–2
Vegoia, *see* 'Begoe'
Velia, 85
Vercellae (Vercelli), 149
Verona, 149
Verrius Flaccus, 86
Verus, surveyor, 69–72
Vespasian, 40–1, 51, 94, 108, 122–3, 157, 160, 175, 177, 184
Vestal Virgins, 186
Veterans, 42, 123
Veturia, 85, 164
Via, *see* Roads
Vibo Valentia, 85
Vicetia, 148
Vienna, 109
Villas, 140–1, 169
Virgil, 31–2, 61, 109, 179, 184
Vitruvius, 39, 48–9, 62, 71, 74, 76, 78–9, 131, 134
Vosges, 141

Walls, 65; Antonine, 199–200
Wetterau, 150
Williamson, H., 204–5
Wolfenbüttel, MSS, 17, 67–8, 101–2, 128–9, 154, 171–2
Woods, 107, 113

York, 195–6